Fibers are ubiquitous. They exist in synthetic and natural forms, as polymers, metals and ceramics. This book is about the processing, microstructure and properties of materials in fibrous form, and is the first to cover the whole range of fibers. Special emphasis is given to major developments in high strength and high stiffness polymeric fibers as well as high temperature ceramic fibers.

The range of fibrous materials covered spans natural polymeric fibers such as silk, synthetic polymeric fibers such as aramid and polyethylene, metallic fibers such as steel and tungsten, and ceramic fibers such as alumina and silicon carbide. The author explains the fundamentals in a clear and concise manner and describes important advances in the production and control of microstructure in high stiffness and high strength fibers. The text contains large numbers of diagrams and micrographs to bring home to the reader the important principles and concepts.

The book will be of value to senior undergraduates, beginning graduate students and researchers in the fields of materials science and engineering, metallurgy, ceramics, textile physics and engineering, mechanical engineering and chemical engineering.

Cambridge Solid State Science Series

Fibrous Materials

TITLES IN PRINT IN THIS SERIES

S.W.S. McKeever
Thermoluminescence of solids

P.L. Rossiter
The electrical resistivity of
metals and alloys

D.I. Bower and W.F. Maddams
The vibrational spectroscopy
of polymers

S. Suresh
Fatigue of materials

J. Zarzycki
Glasses and the vitreous state

R.A. Street
Hydrogenated amorphous silicon

T-W. Chou
Microstructural design of
fiber composites

A.M. Donald and A.H. Windle
Liquid crystalline polymers

B.R. Lawn
Fracture of brittle solids – second
edition

T.W. Clyne and P.J. Withers
An introduction to metal matrix
composites

V.J. McBrierty and K.J. Packer
Nuclear magnetic resonance in
solid polymers

R.H. Boyd and P.J. Phillips
The science of polymer molecules

D.P. Woodruff and T.A. Delchar
Modern techniques of surface
science – second edition

J.S. Dugdale
Electrical properties of metallic glasses

**M. Nastasi, J.M. Mayer
and J.K. Hirvonen**
Ion–solid interactions:
fundamentals and applications

D. Hull and T.W. Clyne
An introduction to composite
materials – second edition

**J.W. Martin, R.H. Doherty
and B. Cantor**
Stability of microstructure in
metallic systems – second edition

**T.G. Nieh, O. Sherby
and J. Wadsworth**
Superplasticity in metals and ceramics

L.V. Gibson and M.F. Ashby
Cellular solids – second edition

Fibrous materials

K. K. Chawla

Department of Materials & Metallurgical Engineering
New Mexico Institute of Mining and Technology

CAMBRIDGE
UNIVERSITY PRESS

PUBLISHED BY THE PRESS SYNDICATE OF THE UNIVERSITY OF CAMBRIDGE
The Pitt Building, Trumpington Street, Cambridge CB2 1RP, United Kingdom

CAMBRIDGE UNIVERSITY PRESS
The Edinburgh Building, Cambridge CB2 2RU, United Kingdom
40 West 20th Street, New York, NY 10011–4211, USA
10 Stamford Road, Oakleigh, Melbourne 3166, Australia

First published 1998

Printed in the United Kingdom at the University Press, Cambridge

Typeset in 10¼/13½pt Monotype Times   [SE]

A catalogue record for this book is available from the British Library

Library of Congress cataloguing in publication data

Chawla, Krishan Kumar, 1942–
    Fibrous materials / K.K. Chawla.
        p.   cm. – (Cambridge solid state science series)
    Includes index.
    ISBN 0 521 57079 4 (hb)
    1. Fibers.   I. Title.   II. Series.
    TA418.9.F5C47   1997
    620.1'1–dc21   96–51098   CIP

ISBN 0 521 57079 4 hardback

एकं सद् विप्रा बहुधा वदंति

The truth is only one;
wise men call it by different names.
                – Rigveda

In loving memory of my dear mother, Sumitra Chawla

Contents

Preface

This book is about materials in fibrous form, precisely what the title says. Perhaps the only thing that needs to be emphasized is that the materials aspects of fibers are highlighted. The main focus is on the triad of processing, microstructure, and properties of materials in a fibrous form. I have kept the mathematics to the bare minimum necessary. More emphasis is placed on physical and chemical insights. Although all kinds of fibers are touched upon, there is a distinct tilt toward synthetic, nonapparel-type fibers. This is understandable inasmuch as the second half of the twentieth century has seen tremendous research and development activity in this area of high performance fibers, mainly for use as a reinforcement in a variety of matrix materials.

The field of fibrous materials is indeed very vast. To compress all the information available in a reasonable amount of space is a daunting task. My aim in writing this text has been to provide a broad coverage of the field that would make the text suitable for anyone generally interested in fibrous materials. I have provided ample references to the original literature and review articles to direct the reader with a special interest in any particular area.

The plan of the book is as follows. After an introductory chapter, some general terms and attributes regarding fibers and products thereof are described in Chapter 2. This chapter also serves to provide a mutually comprehensible language to textile and nontextile users of fibers. There is no gainsaying the fact that many definitions, units and terms about fibers owe their origin to the textile industry. Thus, it behoves a materials scientist or engineer to take cognizance of those and be at home with them. At the same time, it is not unreasonable to expect that a textile engineer should know the stress–strain curves of fibers in engineering units. This general chapter is followed by Chapters 3 and 4 on natural and synthetic polymeric fibers, respectively. Chapter 5 covers metallic fibers, which are quite widely used in a variety of engineering applications, although generally not so recognized. Chapter 6 describes ceramic fibers where much innovative processing work has been done during the last quarter of the twentieth century. This is followed by two chapters (7 and 8) on glass and carbon fibers; two fibers that have been commercially most successful and find widespread use as engineered materials, for example both are used as reinforcements in a variety of composites, and optical glass fiber has an enormous market in telecommunications. Chapter 9 describes some of the testing and characterization techniques used with fibers. Finally, Chapter 10 provides a statistical treatment of strength of fibrous materials. One of the major sources of confusions about fiber characteristics is due to the different units that are used, especially the textile and engineering units. An appendix giving different units and their conversion factors is provided. A book on the materials aspects of fibers, or for that matter any other entity, must have photomicrographs to illustrate the microstructural aspects pertinent to that particular material form and the processing that resulted in that material form. Never was the adage, 'A picture is worth a thousand words', truer than in the present case. I have tried to include as many micrographs as possible. The sources of these are acknowledged in the figure captions.

Acknowledgments

Over the years I have had the good fortune of interacting with a number of extra-ordinary people who have been of great a source of encouragement and help to me in my work. They are (in alphabetical order): M.E. Fine, S.G. Fishman, J.C. Hurt, B. Ilschner, O.T. Inal, S. Kumar, B.A. MacDonald, J.M. Rigsbee, P. Rohatgi, S. Suresh, H. Schneider, N.S. Stoloff, and A.K. Vasudevan. Among my students and post-docs, I would like to thank J. Alba, Jr., B. Furman, A. Neuman, G. Gladysz, J.-S. Ha, and Z.R. Xu. They really made the work seem like fun! No amount of words can convey my debt to Nivi, Nikhil and Kanika. Portions of the book were read by N. Chawla, B. Fureman and S. Kumar. I am grateful to them for their comments. Kanika's help in word processing was invaluable. A special bouquet of thanks to B. Ilschner and J.-A. Manson for their hospitality at Ecole Polytechnique Federale de Lausanne, Switzerland during the summer of 1996. The stay at EPFL provided me with a distraction-free time in very pleasant surroundings to complete this undertaking. Last but not least, I am grateful to my parents, Manohar L. and Sumitra Chawla for all they have done for me!

Chapter 1

Introduction

The term *fiber* conjures up an image of flexible threads, beautiful garments and dresses, and perhaps even some lowly items such as ropes and cords for tying things, and burlap sacks used for transporting commodities, etc. Nature provides us with an immense catalog of examples where materials in a fibrous form are used to make highly complex and multifunctional parts. Protein, which is chemically a variety of complexes of amino acids, is frequently found in nature in a fibrous form. Collagen, for example, is a fibrous protein that forms part of both hard and soft connective tissues. A more well-known natural fiber that is essentially pure protein, is silk fiber. Silk is a very important natural, biological fiber produced by spider and silkworm. It is a solution spun fiber, with the solution, in this case, being produced by the silkworm or the spider. The silkworm silk has been commercialized for many years while scientists and engineers are beginning to realize the potential of spidersilk.

Indeed, materials in a fibrous form have been used by mankind for a long time. Fiber yarns have been used for making fabrics, ropes, and cords, and for many other uses since prehistoric times, long before scientists had any idea of the internal structure of these materials. Weaving of cloth has been an important part of most ancient societies. The term fabric is frequently employed as a metaphor for society. One talks of the social fabric or moral fiber of a society, etc. It is interesting to note that an archeological excavation of a 9000-year-old site in Turkey led to the discovery of a piece of fabric, a piece of linen, woven from the fibers of a flax plant (New York Times, 1993). Normally, archeologists date an era by the pottery of that era. It would appear from this discovery that even before the pottery, there were textile fabrics. There is also recorded use of sutures as stitches in wound repairs in prehistoric times (Lyman, 1991). An ancient medical treatise, about 800 BC, called *The Sushruta Samhita*, written by the Indian surgeon Sushruta, describes the use of braided fibers such as horse hair, cotton fibers, animal sinews, and fibrous bark as sutures.

The importance of fibrous materials in an industrialized economy can hardly be overstated. For example, the fiber related industry is a very large sector of the US economy. According to US fiber industry sources, Americans consume over eight billion (8×10^9) kg fibers per year. In the US, it is an over US\$ 200 billion industry, employs about 12% of the manufacturing workforce, and consumes about 6% of the energy. The fiber industry has about the same importance in Europe, Asia, South America, and other parts of the world. Thus, by any measure, fiber related industrial activity represents a very important sector of the world's economy! Fibrous materials are, in one form or another, part of our daily life. One has only to look at one's surroundings and reflect a little bit to realize the all pervading influence of fibrous materials in our society. From daily uses such as apparel, carpets to artificial turfs, barrier liners under highways and railroad tracks, fiber reinforced composites in sporting goods (rackets, golf shafts, etc.), boats, civil construction, aerospace industry, and defense applications involving aircraft, rocket nozzles and nose cones of missiles and the space shuttle. In short, a quick perusal of our surroundings will show that fibers, albeit in a variety of forms, are used in all kinds of products. A commonplace example that many of us tea drinkers go through everyday involves the use of a tea bag. The proper tea bag needs to be porous, should not impart a taste to the brew, and of course, should have enough wet strength so that it does not fall apart in the hot water. The introduction of the tea bag has an interesting history. Faye Osborne of Dexter Corporation, Windsor Locks, CT, USA is credited with the invention of the tea bag (Sharp, 1995). After trying a number of vegetable fibers, he arrived at wild abaca fiber (or manila hemp) and wood pulp to make the paper for the tea bag. World War II disrupted the availability of manila hemp fiber. In 1942, Osborne came up with rayon fiber made from old rope from which oil had been extracted. A coating of melamine resin helped increase the wet strength of the paper. Of course, there are more sophisticated but less appreciated uses such as fibers for medical uses involving drug delivery, optical fiber that allows examination of inaccessible body parts, etc. Modern usage of fibers in medicine is, of course, quite extensive: surgical dressings and masks, caps, gowns; implantable fibers for sutures, fabrics for vascular grafts and heart repairs; and extra corporeal uses such as fabrics for dialysis and oxygenator membranes, etc. Thus, we live in a world in which matter in a fibrous form is ever present around us.

In this chapter we examine the recent history of synthetic fiber production, provide a convenient classification of fibers, and then introduce the subject of strong and stiff fibers. Strong and stiff fibers came about in the second half of the twentieth century because of many improvements in synthesis and processing,

but most of all, owing to a growing realization of the importance of a processing–structure–property triad. This triad of processing–structure–property
correlations as applied to the fibrous materials is indeed the basic theme of this
book. What we mean by this is that the processing of a material into a fibrous
form determines its microstructure, and the microstructure, in turn, determines
the ultimate properties of the fiber. Time and again we shall come back to this
basic theme in this book.

1.1 Some history

Natural fibers have been around in one form or another from prehistoric times.
Natural silk fiber has been a valuable commodity for a very long time. However,
it was not until about 1880 that a Frenchman, Count Hilaire de Chardonnet
became successful in imitating silkworms and produced the first synthetic fiber
from mulberry pulp. This was *rayon*, not quite silk, but it had the same silky feel
to it. Thus began the era of regenerated cellulosic, natural fibers such as rayon
and acetate. Later, in the mid-1920s synthetic fibers started appearing. The big
breakthrough came when Carothers discovered the process of condensation
polymerization to produce a variety of polymers such as polyamides, polyesters,
and polyurethanes. I think a reasonable case can be made that the modern age
of man-made fibers started with the discovery of the nylon in the 1930s. The
names of two chemists, an American, Wallace Carothers and a German, Paul
Schlack, are linked to the pioneering work that led to the discovery of nylon fiber.
Du Pont Co. started producing nylon fiber in 1939. That can be regarded as the
beginning of the age of man-made fibers; the Second World War, however, inter-
rupted the progress in the development of synthetic fibers. A series of other synthetic fibers was discovered soon after the war, e.g. acrylic, polyester, etc., and
the progress toward the development of synthetic fibers was resumed. It was also
during the second quarter of the twentieth century that scientists started to
unravel the internal microstructure of some of the natural fibers and soon there-
after produced synthetic fibers that rivaled or were improvements on the natural
fibers. Most of this work had to do with applications of fibers for apparel and
similar uses; understandably, therefore, most of the information during the 1940s
and 1950s about fibrous materials came from people involved, in one way or the
other, with textiles. Parallel to the developments in the field of organic fibers
(natural and synthetic), in the late 1930s and early 1940s, it was discovered that
silica-based glass could be drawn into a very high strength fiber. The stiffness of
glass fiber does not differ from that of the bulk glass. This is because glass is an
amorphous material, and therefore, there does not occur any preferential
orientation after the fiber drawing operation. The stiffness of glass is, however,
quite high compared to that of most polymers and therefore glass fiber is quite

suitable for reinforcement of polymeric materials. The advent of glass fibers can be regarded as the harbinger of making and using of fibers in the nontextile domain. Of course, metal wires have been in use for various specific purposes, e.g. copper wire for electrical conduction, tungsten wire for lamp filaments, thermocouple wires made of a variety of alloys for measurement of temperature, steel and other metallic alloy-based wires for a variety of musical instruments such as piano, violin, etc. The last quarter of the twentieth century has seen extensive work in the area of producing high modulus fibers, organic as well as inorganic. In the 1950s, it was realized that the carbon–carbon bond in the backbone chain of polymers is a very strong one and that if we could only orient the molecular chains, we should get a high modulus fiber from the organic fibers. This was attempted, at first, by applying ever increasing stretch or draw ratios to the spun organic fibers. This resulted in some improvement in the stiffness of the fiber, but no better than what was available with glass fiber, i.e. a tensile or Young's modulus of about 70 GPa in the fiber direction. In time, however, researchers realized that what one needed was orientation *and* extension of the molecular chains in order to realize the full potential of the carbon–carbon bond. Aramid and ultra-high molecular weight polyethylene fibers, with Young's modulus over 100 GPa, epitomize this oriented and extended chain structure. The last quarter of the twentieth century also saw an increasing amount of work in the area of ceramic fibers having low density, high stiffness, high strength, but, more importantly, possessing these characteristics at high temperatures (as high as 1000°C and over). Examples of such high temperature fibers include boron, carbon, alumina, silicon carbide, etc. Some very novel and innovative techniques based on sol–gel and the use of polymeric precursors to obtain inorganic compounds or ceramic fibers such as silicon carbide, alumina, etc. came into being. These very significant advances opened up an entirely new chapter in the field of fibrous materials, namely fibers that have very high elastic stiffness, high strength, low density, and are capable of withstanding extremely high temperatures. The driving force for this development was the use of these fibers as reinforcements for metals and ceramics, i.e. at medium to very high temperatures.

1.2   Classification of fibers

One can classify fibers in a variety of ways. For example, one may divide the whole field of fibers into apparel and nonapparel fibers, i.e. based upon the final use of fibrous material. The apparel fibers include synthetic fibers such as nylon, polyester, spandex, and natural fibers such as cotton, jute, sisal, ramie, silk, etc. Nonapparel fibers include aramid, polyethylene, steel, copper, carbon, glass, silicon carbide, and alumina. These nonapparel fibers are used for making cords and ropes, geotextiles, and structural applications such as fiber reinforcements

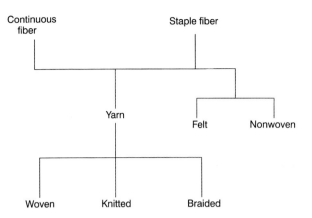

Figure 1.1 Classification of fibers based on fiber length. Also shown are the different product forms.

in a variety of composites. These fibers possess higher stiffness and strength and a lower strain to failure than the apparel fibers. They are also characterized by rather difficult processing and a drastic strength degradation by the presence of small flaws, i.e. they generally have a low toughness. Another classification of fibers can be made in terms of fiber length, continuous or staple fiber. Continuous fibers have practically an infinite length while staple fibers have short, discrete lengths (10–400 mm). Like continuous fibers, these staple fibers can also be spun into a yarn, called staple fiber yarn (see Chapter 2). This ability to be spun into a yarn can be improved if the fiber is imparted a waviness or crimp. Staple fibers are excellent for providing bulkiness for filling, filtration, etc. Frequently, staple natural fibers (cotton, wool) and staple synthetic fibers (nylon, polester) are blended to obtain the desirable characteristics of both. Figure 1.1 shows this classification based on fiber size, and the different product forms that are commonly available, e.g. woven or nonwoven. Yet another convenient classification of fibers is based on natural and synthetic fibers, as shown in Fig. 1.2. Natural fibers occurring in the vegetable or animal kingdom are polymeric in terms of their chemical constitution, while natural fibers in the form of minerals are akin to crystalline ceramics. A distinctive feature of natural fibers is that they are generally a mixture (chemical or physical) of different compounds. One can further classify synthetic fibers as polymers, metals, and ceramics or glass. Here, one should also mention a very special and unique subclass of fibers, *whiskers*. Whiskers are monocrystalline, short fibers with extremely high strength. This high strength, approaching the theoretical strength, comes about because of the absence of crystalline imperfections such as dislocations. Being monocrystalline, there are no grain boundaries either. Whiskers are normally obtained by vapor phase growth. Typically, they have a diameter of a few micrometers and a length of a few millimeters. Thus, their aspect ratio (length/diameter) can vary from 50 to 10 000. Perhaps the greatest drawback of whiskers is that they do not have uniform dimensions or properties. This results

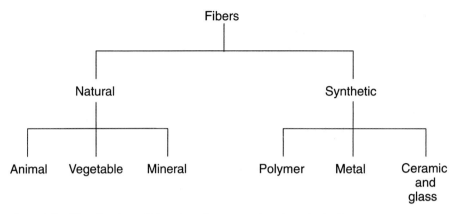

Figure 1.2   Classification of fibers based on natural and synthetic fibers.

in an extremely large spread in their properties. Handling and alignment of whiskers in a matrix to produce a composite are other problems.

1.3   **Stiff and strong fibers**

As we pointed out earlier, fibrous material is found extensively in nature. Up until the mid-twentieth century, most of the usage of fibers had been in clothing and other household uses. About the middle of the twentieth century, high performance fibers became available for use in a fabric form or as reinforcements for making composites. Our main focus in this text will be on processing, microstructure, properties, and applications of *materials* in a fibrous form, with a distinct emphasis on synthetic fibers for nontextile applications. Such fibers are generally very stiff and strong. This is where some of the important developments have occurred in the second half of the twentieth century. The use of fibers as high performance materials in structural engineering applications is based on three important characteristics:

(i) A small diameter with respect to its grain size or other microstructural unit. This allows a higher fraction of the theoretical strength to be attained than that possible in a bulk form. This is a direct result of the so-called size effect, namely the smaller the size, the lower the probability of having an imperfection of a critical size that would lead to failure of the material. Thus, even for a material in its fibrous form, its strength decreases as its diameter increases.

(ii) A very high degree of flexibility which is really a characteristic of a material having a high modulus and a small diameter. This flexibility permits a variety of techniques to be employed for making fabrics, ropes, cords, and fiber reinforced composites with these fibers.

(iii) A high aspect ratio (length/diameter, l/d) which allows a very large fraction of applied load to be transferred via the matrix to the stiff and strong fiber in a fiber reinforced composite (Chawla, 1987)

The most distinctive feature of a fibrous material is that it has properties highly biased along its length. Fibers, in general, and continuous fibers, in particular, are very attractive for the reasons given above. A given material in a fibrous form has a high aspect ratio (length/diameter) and can be highly flexible. Such a flexible fiber can be made into yarn, which in turn can be braided, knitted, or woven into rather complex shapes and forms. Think of some materials that are inherently brittle in their bulk form such as glass, alumina, or silicon carbide, etc. In the form of an ultrafine diameter fiber, they can be as flexible as any organic textile fiber, such as nylon that is commonly used to make women's stockings. Quite frequently, as we mentioned earlier, a material in a fibrous form has a higher strength and, sometimes, as in a highly oriented fiber, even a higher elastic modulus than in the bulk form. These characteristics have led to the development of fiber reinforced composites with a variety of matrix materials such as polymers, metals, glasses, and ceramics (see, for example, Chawla, 1987, 1993; Chou and Ko, 1989; Hull and Clyne, 1996; Piggott, 1980; Taya and Arsenault, 1989; Suresh et al., 1993; Clyne and Withers, 1993). In the chapters to follow we describe the processing–microstructure–properties of some of these fibrous materials.

Chapter 2

Fibers and fiber products

In this chapter, we define some important terms and parameters that are commonly used with fibers and fiber products such as yarns, fabrics, etc., and then describe some general features of fibers and their products. These definitions, parameters, and features serve to characterize a variety of fibers and products made from them, *excluding* items such as fiber reinforced composites. These definitions and features are generally independent of fiber type, i.e. polymeric, metallic, glass or ceramic fibers. They depend on the geometry rather than any material characteristics.

Fiber is the fundamental unit in making textile yarns and fabrics. Fibers can be naturally occurring or synthetic, i.e. man-made. There are many *natural fibers*, organic and inorganic. Examples include organic fibers such as silk, wool, cotton, jute, sisal and inorganic fibers such as asbestos and basalt. There is a large variety of *synthetic fibers* available commercially. Polymer fibers such as polypropylene, polyethylene, polyamides, polyethyleneterephthalate (PET), polyacrylonitrile (PAN), polytetrafluoroethylene (PTFE), aramid, etc. are well-established fibers. Metallic wires or filaments have been available for a long time. Examples include steel, aluminum, copper, tungsten, molybdenum, gold, silver, etc. Among ceramic and glass fibers, glass fiber for polymer reinforcement has been available since the 1940s; optical glass fiber for telecommunication purposes made its debut in the 1950s, while ceramic fibers such as carbon, silicon carbide, alumina, etc. became available from the 1960s onward.

One can transform practically any material, polymer, metal, or ceramic, into a fibrous form. As we pointed out in Chapter 1, historically and traditionally, fibers formed part of the textile industry domain for uses such as clothing, upholstery and draperies, sacks, ropes, cords, sails, and containers, etc. Gradually, their use entered the realm of more engineered items such as conveyor belts, drive belts, geotextiles, etc. With the advent of high modulus fibers, the use of fibers has extended to highly engineered materials such as composites. Understandably, therefore, some of the terms commonly used with fibers have their origin in textile technology terminology. We examine some of this terminology next. This will prepare some common ground among people interested in fibers, in one way or another, but who come from different disciplines.

2.1   Definitions

We first define some terms commonly used in the field of fibrous materials. We should add that some of these definitions are expanded upon later in this chapter. However, before one can define the term fiber, one needs to define the most important attribute of the fiber that serves to define a fiber, namely the fiber aspect ratio. The aspect ratio of a fiber is the ratio of its length to diameter (or thickness). A *fiber* can be defined as an elongated material having a more or less uniform diameter or thickness less than 250 μm and an aspect ratio greater than 100. This is an operational definition of fiber. It is also a purely geometrical one in that it applies to any material. Having defined the basic unit, the fiber, we are now in a position to define some other commonly used terms related to fibers. These are given below in alphabetical order:

- *Aspect ratio:* The ratio of length to diameter of a fiber.
- *Bicomponent fibers:* A fiber made by spinning two compositions concurrently in each capillary of the spinneret.
- *Blend:* A mix of natural staple fibers such as cotton or wool and synthetic staple fibers such as nylon, polyester. Blends are made to take advantages of the natural and synthetic fibers.
- *Braiding:* Two or more yarns are intertwined to form an elongated structure. The long direction is called the bias direction or machine direction.
- *Carding:* Process of making fibers parallel by using rollers covered with needles.
- *Chopped strand:* Fibers are chopped to various lengths, 3 to 50 mm, for mixing with resins.
- *Continuous fibers:* Continuous strands of fibers, generally, available as wound fiber spools.
- *Cord:* A relatively thick fibrous product made by twisting together two or more plies of yarn.

- *Covering power:* The ability of fibers to occupy space. Noncircular fibers have a greater covering power than circular fibers.
- *Crimp:* Waviness along the fiber length. Some natural fibers, e.g. wool, have a natural crimp. In synthetic polymeric fibers crimp can be introduced by passing the filament between rollers having teeth. Crimp can also be introduced by chemical means. This is done by controlling the coagulation of the filament to produce an asymmetrical cross-section.
- *Denier:* A unit of linear density. It is the weight in grams of 9000 m long yarn. This unit is commonly used in the US textile industry.
- *Fabric:* A kind of planar fibrous assembly. It allows the high degree of anisotropy characteristic of yarn to be minimized, although not completely eliminated.
- *Felt:* Homogeneous fibrous structure made by interlocking fibers via application of heat, moisture, and pressure.
- *Filament:* Continuous fiber, i.e. fiber with aspect ratio approaching infinity.
- *Fill:* see *Weft.*
- *Handle:* Also known as softness of handle. It is a function of denier (or tex), compliance, cross-section, crimp, moisture absorption, and surface roughness of the fiber.
- *Knitted fabric:* One set of yarn is looped and interlocked to form a planar structure.
- *Knitting:* This involves drawing loops of yarns over previous loops, also called interlooping.
- *Mat:* Randomly dispersed chopped fibers or continuous fiber strands, held together with a binder. The binder can be resin compatible, if the mat is to be used to make a polymeric composite.
- *Microfibers:* Also known as microdenier fibers. These are fibers having less than 1 denier per filament (or less than 0.11 tex per filament). Fabrics made of such microfibers have superior silk-like handle and dense construction. They find applications in stretch fabrics, lingerie, rain wear, etc.
- *Monofilament:* A large diameter continuous fiber, generally, with a diameter greater than 100 μm.
- *Nonwovens:* Randomly arranged fibers without making fiber yarns. Nonwovens can be formed by spunbonding, resin bonding, or needle punching. A planar sheet-like fabric is produced from fibers without going through the yarn spinning step. Chemical bonding and/or mechanical interlocking is achieved. Fibers (continuous or staple) are dispersed in a fluid (i.e. a liquid or air) and laid in a sheet-like planar form on a

support and then chemically bonded or mechanically interlocked. Paper is perhaps the best example of a wet-laid nonwoven fabric where we generally use wood or cellulosic fibers. In spunbonded nonwovens, continuous fibers are extruded and collected in random planar network and bonded.

- *Particle:* Extreme case of a fibrous form: it has a more or less equiaxial form, i.e. the aspect ratio is about 1.
- *Plaiting:* see *Braiding.*
- *Rayon:* Term used to designate any of the regenerated fibers made by the viscose, cuprammonium, or acetate processes. They are considered to be natural fibers because they are made from regenerated, natural cellulose.
- *Retting:* A biological process of degrading pectin and lignin associated with vegetable fibers, loosening the stem and fibers, followed by their separation.
- *Ribbon:* Fiber of rectangular cross-section with width to thickness ratio greater than 4.
- *Rope:* Linear flexible structure with a minimum diameter of 4 mm. It generally has three strands twisted together in a helix. The rope characteristics are defined by two parameters, unit mass and break length. Unit mass is simply g/m or ktex, while breaking length is the length of rope that will break under the force of its own weight when freely suspended. Thus, break length equals mass at break/unit mass.
- *Roving:* A bundle of yarns or tows of continuous filaments (twisted or untwisted).
- *Spinneret:* A vessel with numerous shaped holes at the bottom through which a material in molten state is forced out in the form of fine filaments or threads.
- *Spunbonding:* Process of producing a bond between nonwoven fibers by heating the fibers to near their melting point.
- *Staple fiber:* Fibers having short, discrete lengths (10–400 mm long) that can be spun into a yarn are called staple fibers. This spinning quality can be improved if the fiber is imparted a waviness or crimp. Staple fibers are excellent for providing bulkiness for filling, filtration, etc. Frequently, staple natural fibers, e.g. cotton or wool, are blended with staple synthetic fibers, e.g. nylon or polyester, to obtain the best of both types.
- *Tenacity:* A measure of fiber strength that is commonly used in the textile industry. Commonly, the units are gram-force per denier, gram-force per tex, or newtons per tex. It is a specific strength unit, i.e. there is a factor of density involved. Thus, although the tensile strength of glass fiber is more than double that of nylon fiber, both glass and nylon fiber have a

tenacity of about 6 g/den. This is because the density of glass is about twice that of nylon.

- *Tex:* A unit of linear density. It is the weight in grams of 1000 m of yarn. Tex is commonly used in Europe.
- *Tow:* Bundle of twisted or untwisted continuous fibers. A tow may contain tens or hundreds of thousands of individual filaments.
- *Twist:* The angle of twist that individual filaments may have about the yarn axis. Most yarns have filaments twisted because it is easier to handle a twisted yarn than an untwisted one.
- *Wire:* Metallic filament.
- *Warp:* Lengthwise yarn in a woven fabric.
- *Weft:* Transverse yarn in a woven fabric. Also called *fill.*
- *Whisker:* Tiny, whisker-like fiber (a few mm long, a few μm in diameter) that is a single crystal and almost free of dislocations. Note that this term involves a *material* requirement. The small size and crystalline perfection make whiskers extremely strong, approaching the theoretical strength.
- *Woven fabric:* Flat, drapeable sheet made by interlacing yarns or tows.
- *Woven roving:* Heavy, drapeable fabric woven from continuous rovings.
- *Yarn:* A generic term for a bundle of untwisted or twisted fibers (short or continuous). A yarn can be produced from staple fibers by yarn spinning. The yarn spinning process consists of some fiber alignment, followed by locking together by twisting. Continuous synthetic fibers are also used to make yarns. Continuous fibers are easy to align parallel to the yarn axis. Generally, the degree of twist is low, just enough to give some interfilament cohesion.

2.1.1   Yarn

Fibers can be made into a variety of very useful product forms. Most of these can be classified under the categories of woven or nonwoven. Most of the woven product forms are based on fibrous yarns. Next we describe some of these product forms involving yarn.

Fibers are frequently used in the form of a yarn, mainly because a multifilament yarn is more flexible and pliable than a solid monofilament of the same diameter. The process of making a continuous, multifilament thread or yarn from continuous or staple fibers is called *spinning*. This nomenclature is rather unfortunate because the same term is also used to denote the process of making individual fibers by extruding through a spinneret. The fibers in a yarn are held in place through radial frictional compressive force generated by the twist that

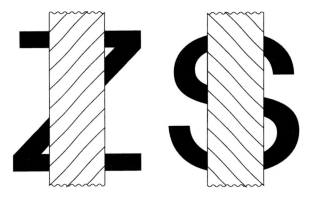

Figure 2.1 Two types of yarn twists: a clockwise twist or Z twist and a counterclockwise twist or S twist.

is given to the yarn. When such a twisted yarn is pulled in tension, the individual filaments are forced together and the load is shared evenly by the multifilaments in the yarn. The frictional force increases with the angle of twist.

A fiber yarn is a multifilament assembly. A yarn is frequently employed in making woven forms. There are certain yarn structural parameters that are important in determining its properties. Some of the important yarn structural parameters are described below.

Linear density: This gives us an idea of the yarn size. *Denier* is the weight in grams of a 9000 m length of yarn. This is commonly used in the US. *Tex* is the weight in grams of 1000 m of yarn. Tex is commonly used in Europe. Strictly speaking, these linear density terms can be applied to single fibers as well. The basic idea is that the higher the linear density (denier or tex), the heavier the yarn.

For example, an individual fiber may be 1.5 den, while a yarn made of a collection of the same fibers may be available in 200 den, 400 den, and all the way to 1500 den. Clearly, as the denier, or tex, increases, the larger is the number of individual fibers in the yarn.

Number of fibers: This gives the number of individual filaments in the yarn. This parameter is really fixed by the number of holes in the spinneret.

Twist in the yarn: Most yarns have filaments that are twisted. The main reason for this is that an untwisted yarn is difficult to weave or knit. Two types of twists can be given to the yarn, a counterclockwise twist or S twist and a clockwise twist or a Z twist. Figure 2.1 shows these twists. We can also make a ply yarn by using reverse twist directions. This serves to balance out residual stresses. We can also twist together two or more plies to make a cord. Commonly, yarn designation on a fiber spool provides information such as name, linear density, number of fibers and fiber type.

Next we describe some general features of the fiber spinning process and some useful but very general characteristics of fibers. These features allow fibers to be made into yarns which can be woven, braided or knitted to make rather diverse and useful structural forms. We give below a summary of these.

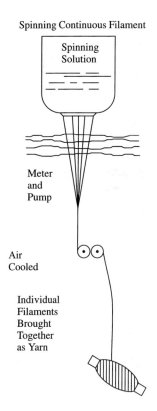

Spinning Continuous Filament

Spinning
Solution

Meter
and
Pump

Air
Cooled

Individual
Filaments
Brought
Together
as Yarn

Figure 2.2   Schematic of a generic fiber spinning process.

2.1.2   Fiber spinning

The process of producing a fibrous form from the liquid state is called spinning. There are quite a few variants of the basic fiber spinning process. The most common is called melt spinning, which as the name implies means that the fiber is produced from a melt. This process is used for producing fibers from organic polymers and inorganic silica-based glasses. An appropriate thermoplastic polymer, generally in the form of dry pellets, is fed into an extruder where it is melted and then pumped through a spinneret to form filaments. A spinneret has multiple holes through which the molten material exits as fine filaments. Figure 2.2 shows a schematic of a generic fiber spinning process. We shall describe the detailed variants of this process in subsequent chapters. Suffice it here to mention three important types of polymeric fiber spinning processes: dry spinning, wet spinning and dry jet–wet spinning. In the dry spinning process the polymer melt or solution is extruded into an evaporating gaseous stream. In the wet spinning process molten jets are extruded into a precipitating liquid medium. In dry jet–wet spinning the extruded solution passes through an air gap before entering a coagulation bath.

The following process variables are very important in any fiber spinning process:

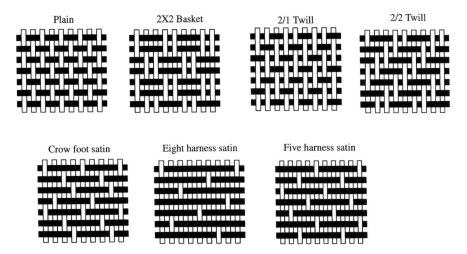

Plain   2X2 Basket   2/1 Twill   2/2 Twill

Crow foot satin   Eight harness satin   Five harness satin

Figure 2.3   Basic weaves.

- flow rate of the melt;
- take-up velocity of the wound filaments;
- extrusion temperature;
- dimensions and shape of the spinneret holes;
- cooling conditions.

We describe the detailed spinning process of different individual fibers in their respective chapters.

2.1.3   Weave

Weaving is a common method of making a planar fabric. Some of the important weave patterns used to make woven fabrics are shown in Fig. 2.3. *Warp* yarns are the lengthwise fibers, while the *fill or weft* yarns are in the transverse direction. Figure 2.3 shows these warp and fill (or weft) directions in two types of weaves: plain and satin. A plain weave has warp ends going alternately over and under a fill yarn (or weft) yarn. A plain weave provides a firm construction, i.e. very little slippage occurs. The tensile strength of the fabric is high and uniform in the two orthogonal directions, but its in-plane shear strength is rather poor. Such a fabric has high porosity. In the satin weave, warp ends go over a certain number of fill yarns and then under one fill yarn. For example, in the five-harness satin, shown in Fig. 2.3, warp ends go over, successively, four fill yarns and then under one. A satin weave is more pliable or drapeable than a plain weave and can take complex shapes and contours. It is also less porous than a plain weave.

It should be clear that the characteristics of a fabric depend on the characteristics of the fibrous materials used as well as the geometry of the yarn and weave.

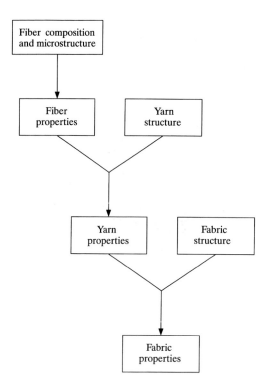

Figure 2.4 Flow chart for obtaining a fabric with specific characteristics starting from a fiber.

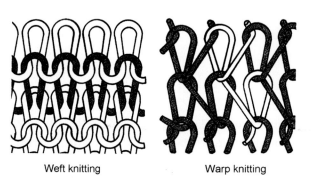

Weft knitting   Warp knitting

Figure 2.5 Interlooping of a yarn (after Mohamed, 1990).

Figure 2.4 summarizes the process of obtaining a fabric with specific characteristics starting from a fiber.

2.1.4 Knitting

Knitting is another way of producing a fabric. This process involves interlooping one set of yarns, i.e. drawing loops of one yarn over previous loops. Figure 2.5 shows the interlooping of a yarn in two different ways: weft knitting and warp knitting (Mohamed, 1990). One great advantage of a knitted fabric is high extensibility in all directions: think of putting a knitted sweater over your head!

Figure 2.6 Schematic of a braided fabric structure in relaxed and stressed conditions.

Relaxed

Stressed

Different knitting patterns can be tailored to requirements of directional elongation and stability. The disadvantages of this process of making a fabric include the following:

(a) The pores in knitted fabric are not of a uniform size and shape.

(b) Knitted fabrics have relatively large thickness, i.e. the yarn consumption per unit area is high.

2.1.5 Braiding

Braiding involves interlacing of a yarn in what is called the bias direction. Figure 2.6 shows a braided fabric structure. It is similar to a woven fabric rotated

Figure 2.7 Braiding of a sleeve in progress (courtesy of Atkins & Pierce Co.).

through 45°. A biaxially braided fabric has a high torsional stability but is highly deformable under tension. A triaxial braiding has longitudinal yarns in addition to the two in biaxial braiding. This enhances its tensile and compressive strength.

Yarns can be braided to make a variety of useful products. For example, braiding is a common technique to make sleevings of carbon, aramid, glass, ceramic fibers for use as reinforcement in a variety of matrix materials. Braiding of commingled and hybrid yarns (say, carbon and glass) to make flat or tubular products is quite common. Typical examples of the use of fiber reinforcements can be found in aerospace and defense (weapon systems), industrial uses such as automotive and electric wire sleeving, and recreational equipment parts such as tubular parts used in a bicycle. A new 'mandrel-less' braiding process has been developed to produce braided preforms for fishing rods and golf club shafts. Figure 2.6 shows schematically a braided piece in relaxed and stressed conditions. A picture of actual braiding being done is shown in Fig. 2.7. A comparison of braiding, weaving, and knitting techniques is presented in Table 2.1 (Ko, 1987).

2.1.6   Nonwovens

Nonwovens is a general term used to denote fibrous products that are made without using the weaving, knitting or braiding processes described above.

Nonwoven fibrous products are cheaper to produce because the number of steps required is small. Weaving, on the other hand, can be very complex and consequently expensive. Nonwoven products such as mats are porous and have lower strength and flexibility compared to the woven fabrics. Nonwovens can be made by a variety of different methods:

- *Thermally bonded*: Randomly laid fibers are held together by thermal fusion at myriad points.
- *Resin bonded*: A polymeric resin or glue is introduced and the fiber webs are held together by this glue or resin.
- *Needle punched*: Punched needles are introduced to entangle the webs of fibers.

Nonwoven fibrous products are characterized by assemblies of fibers lying in a plane and bonded at random points. Figure 2.8 shows the microstructure of nonwoven mats of glass and carbon fibers. Crimp in the fibers, natural or artificially created, helps in interlocking the fibers. Examples of nonwovens include disposable diapers, paper, paper towels, napkins, burn dressings, a variety of filter and insulation products, and felts. A felt is a particularly good example of a nonwoven fibrous structure. Felt-like nonwovens can easily be formed into a variety of shapes by merely compressing. Examples of felt products include mats, blankets, carpets, hats, a variety of insulation products, etc. Thick felt-like structures are used as geotextile underlays for roads and railroad beds. Felt and felt products can be made from virtually any type of fibers (Anderson, 1993). Natural fibers such as wood pulp, fur, and hair have scales and sometimes a natural crimp that make interlocking of fibers easy. As mentioned above, generally, heat, pressure, moisture, and agitation are used to make fairly dense felt products. Wool in particular is so well suited for felt making that it is frequently blended (as little as 10%) with 'nonfelting' synthetic or natural fibers to produce a felting mat. Felts can have areal weights ranging from 100 to 3600 g m^{-2} and thicknesses ranging from 0.5 to 7.5 mm. Generally, no adhesives are used in making felts.

Table 2.1 *A comparison of braiding, weaving, and knitting techniques (after Ko, 1987).*

	Braiding	Weaving	Knitting
Direction of yarn	One	Two (0°/90°)	One (0° or 90°)
Formation technique	Intertwining (position displacement)	Interlacing (insert 90° yarns into a 0° yarn system)	Interlooping (drawing loops of yarn over previous loops)

Figure 2.8
Microstructure of non-woven mats of (a) glass and (b) carbon fibers.

Carding

Nonwoven webs can be made by a process called carding which involves mechanically opening, combing, and aligning staple fibers (Crook, 1993). Rotating cylinders or plates covered with a metal wire are used to make carded webs. The process results in nonwovens containing quite uniform fiber distribution. The cardability of a staple fiber depends on its geometry, diameter, length, frictional characteristics, crimp, etc.

2.1.7 Rope

Rope represents a very useful form of fibrous product. A rope or cord consists of a bundle of fibers. The fibers may be continuous or they may be made of staple fibers, i.e. short, fine fibers. The tensile strength of a rope comes from the strength of individual fibers and the friction between them. The interfiber friction prevents their slip past one another. Quite obviously, a rope or a cord has very anisotropic properties. It is strong in tension along the axis direction but not in the transverse or radial direction. Strength in compression is also very poor.

2.2   Some important attributes of fibers

A material in fibrous form has a series of attributes characteristic of its fibrous state. Some of these are obvious, e.g. the properties of a fiber are highly biased in the fiber direction, while others are not so obvious. Some of these characteristics stem mainly from their small cross-section and large aspect ratio, for example:

- a high degree of flexibility;
- a higher strength than the bulk material of the same composition.

These characteristics of fibers lead to some unique features which must be considered in any product that uses fibrous materials. In this section, we describe some of these features that stem from its fibrous nature.

2.2.1 Fiber length per unit weight or volume

We can readily obtain some simple geometric relationships. For example, fiber length per unit weight or volume. We can write fiber volume v as

$$v = \pi d^2 \ell / 4$$

where d is the fiber diameter and ℓ is the fiber length. Rearranging, we can write

$$\ell = 4v / \pi d^2 \tag{2.1}$$

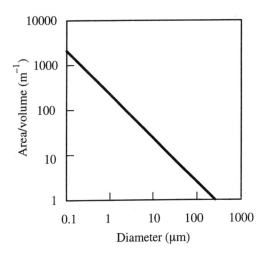

Similarly, if we consider the mass of the fiber, m, rather than its volume, we can write

$$\ell = 4m/\rho\pi d^2 \tag{2.2}$$

where ρ is the material density.

Both of these simple expressions (Eqs (2.1) and (2.2)) tell us that for a given mass or volume of fiber, its length varies inversely as the square of the fiber diameter. Thus, if we halve the fiber diameter, we will increase its length fourfold.

2.2.2 Fiber surface area

A given material in a fibrous form can have a very large surface area. Consider a fiber of length ℓ and diameter d. Then, for the fiber volume, we can write

$$v_f = \text{volume of fiber} = \frac{\pi}{4}d^2\ell \tag{2.3}$$

The fiber surface area, S_A, can be expressed as

$$S_A = \pi d\ell \tag{2.4}$$

From Eqs (2.3) and (2.4), we obtain

$$\frac{S_A}{v_f} = \frac{4}{d} \tag{2.5}$$

Equation (2.5) says that the fiber surface area, for a given fiber volume, varies inversely as the fiber diameter. Figure 2.9 shows a plot of the fiber surface area per unit volume as a function of fiber diameter. As one goes from the large diameter CVD type fibers such as SiC, B, etc., to extremely small fibers such as carbon, aramid, alumina, etc., the fiber surface area per unit volume increases. This has very important implications in composite materials because the fiber/matrix interface has a crucial role in determining the ultimate properties of a composite (Chawla, 1987, 1993). For a given fiber volume fraction in a composite, the fiber surface area will determine the fiber/matrix interfacial area. Equation (2.5) tells us that as the fiber diameter decreases, the fiber surface increases, and, consequently, the interfacial area in a fibrous composite will increase.

2.2.3 Fiber crimp

Crimp is a term used to describe the waviness of a fiber. We can describe the crimp of a fiber in purely geometric terms such as wave amplitude or frequency. We can also describe this in terms of the force or energy required to uncrimp a fiber. Crimp in a fiber is desirable because it makes it easy to process the fiber into a variety of forms such as a spun yarn. Commonly, one encounters such waviness in protein-based animal fibers, such as human hair, and especially wool. Figure 2.10 shows a scanning electron microscopy (SEM) micrograph of a human hair where scales can be seen. In the synthetic polymeric fibers such crimp can be introduced by passing the filament between rollers having teeth. Crimp can also be introduced by chemical means. This is done by controlling the coagulation of the filament to produce an asymmetrical cross-section. Another chemical means of introducing crimp is to have a bicomponent fiber. Miraflex fiber is such a fiber. It is a glass fiber with different chemical compositions in two halves of its cross-sections (see Chapter 7). Fibers having short, discrete lengths that can be spun into a yarn are called staple fibers. Staple fibers are, generally, spun into a long yarn suitable for subsequent textile processing. The spinning quality of staple fibers can be improved if the fiber is imparted a waviness or crimp.

2.2.4 Fiber flexibility

Flexibility of a fiber, i.e. its ability to be bent to an arbitrary radius is one of the important attributes of a fibrous material and worth discussing in some detail. It is easy to visualize that many operations such as weaving, braiding, winding, etc., depend on the ability of a fiber to be bent without breaking. This flexibility is also of great importance in composites because it permits a variety of techniques to be employed for making composites with these fibers. If we treat the fiber as an elastic beam of circular cross-section, then we can easily see that the fiber flexibility corresponds to the ability of the elastic beam to be bent to an

Figure 2.10 SEM micrograph of a human hair showing scales.

arbitrary radius of curvature. A high degree of flexibility is really a characteristic of a material having a low modulus and a small diameter (Dresher, 1969). The flexibility of a given fiber is a function of its stiffness or elastic modulus, E, and the moment of inertia of its cross-section, I. The elastic modulus of a material is quite independent of its form or size. It is generally a material constant for a given chemical composition and fully dense material. Thus, for a given composition and density, the flexibility of a material is determined by its shape, the size of its cross-section, and its radius of curvature, which is a function of its strength. We can use the inverse of the product of bending moment (M) and the radius of curvature (R) as a measure of flexibility. From elementary strength of materials, we have the following relationship for a beam:

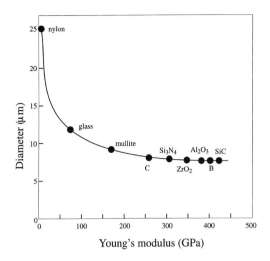

Figure 2.11 Diameter of various fibers with a flexibility equal to that of a 25 μm diameter nylon fiber. Given a sufficiently small diameter, it is possible to produce, in principle, an equally flexible fiber from a polymer, a metal, or a ceramic.

$$M/I = E/R$$

Rearranging and remembering that the moment of inertia, I, for a beam of circular cross-section of diameter d, is equal to $\pi d^4/64$, we can write

$$MR = EI = E\pi d^4/64$$

Or, we can write for the flexibility of a fiber,

$$1/MR = 64/E\pi d^4 \approx 20/Ed^4 \qquad (2.6)$$

Equation (2.6) indicates that $(1/MR)$, a measure of flexibility, is a very sensitive function of diameter, d. If we take a 25 μm diameter nylon fiber as a quintessential example of a flexible fiber, we can compute the diameter of various other fibers that will be required to have a flexibility equal to that of a 25 μm diameter nylon fiber. Figure 2.11 shows this. It follows from this curve that, given a sufficiently small diameter, it is possible to produce, in principle, an equally flexible fiber from a polymer, a metal, or a ceramic. In other words, one can make very flexible fiber out of a ceramic such as silicon carbide or alumina provided one can make it into a fine enough diameter. Making a fine diameter ceramic fiber, however, can be a formidable problem in ceramic processing.

2.2.5   Skin–core structure

As we shall see in the chapters to follow, melt or solution spinning is perhaps the most common method of giving a fibrous form to a material. Even in the case of ceramic fibers that are produced from polymeric precursor fibers, high speed

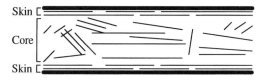

Figure 2.12   Schematic of a skin core structure in a high speed spun fiber showing a greater degree of crystallization in the skin than in the core.

spinning is used to produce the precursor fiber. This high speed spinning process can have some important effects on the resultant fiber structure and, consequently, on its properties. The process of extrusion through a spinneret leads to a certain chain orientation in the filament. Generally, the molecules in the surface region undergo more orientation than the ones in the interior because the solution in contact with the sides of the spinneret hole suffers more frictional resistance than the solution in the center of the spinneret hole. In other words, high speed spun fibers show a skin–core structure, i.e. there are radial variations in the microstructure of a fiber. Microstructurally, these variations correspond to a higher degree of molecular orientation and crystallinity in the near-surface region (skin region) than in the core region of the fiber (Perez, 1985; Vassilatos *et al.*, 1985). Another reason for this skin effect is the presence of a radial temperature variation (lower temperature in the near surface region than in the interior) due possibly to cross flow of the quench air. Such a radial temperature variation results in a higher viscosity and stress in the skin region than in the core. Consequently, most high speed spun fibers show a greater degree of crystallization in the skin than in the core. Such a skin–core structure is shown schematically in Fig. 2.12. This skin effect can affect many other properties of the fiber, for example, the adhesion with a polymeric matrix or the ability to be dyed.

2.2.6   Stretching and orientation

As pointed out above, the spinning process of making a fiber results in some chain orientation but this occurs concomitant with a skin effect. Generally, the as-spun polymeric fiber is subjected to some stretching, causing further chain orientation along the fiber axis and consequently better tensile properties, such as stiffness and strength along the fiber axis. The amount of stretch is generally described in terms of a draw ratio which is the ratio of the initial diameter to the final diameter. For example, nylon fibers are typically subjected to a draw ratio of 5 after spinning. The higher the draw ratio, the higher the degree of chain alignment, the higher the degree of crystallinity, and consequently, the higher will be the elastic modulus of the fiber along its axis. This can affect a number of fiber characteristics, for example, the ability of a fiber to absorb moisture. The higher the degree of crystallinity, the lower the moisture absorption. A good example of such a behavior is that of aramid fiber (see Chapter 4). Du Pont produces a series of aramid fibers under the trade name of Kevlar. In particular, Kevlar 149 fiber has a higher degree of crystallinity than Kevlar 49. Consequently, Kevlar 149 has a higher elastic modulus and absorbs less moisture than Kevlar 49. In general, one can say, that a higher degree of crystallinity in a polymeric fiber translates into a higher resistance to penetration by foreign molecules. This means a greater chemical stability. This can have important effects on textile fibers as it affects the dyeing characteristics of fiber. A high degree of crystallinity makes it difficult for the dye molecules to penetrate the fiber.

There is, however, a limit to the amount of stretch that can be given to a polymer because the phenomenon of necking can intervene and cause rupture of the fiber. In other words, in a polymeric fiber there is a limit to the modulus enhancement that can be obtained by subjecting it to an ever higher draw ratio. As we shall see in Chapter 4, this led to other means of obtaining high stiffness polymeric fibers such as aramid and polyethylene.

2.2.7 Cross-sectional shape and surface roughness

The cross-sectional shape of a fiber can affect many properties, e.g. luster, density, optical properties, feel of the fabric and an important characteristic called the covering power of a fabric. Covering power is the ability of fibers to occupy space. The reader can easily visualize that fibers having a circular cross-section will have a lesser covering power than fibers having a lobed cross-section. Noncircular fibers can provide a greater density in a fabric than circular fibers.

Fibers can have variety of cross-sectional shapes, although most fibers have a more or less circular cross-sectional shape (see Fig. 2.13). Natural fibers mostly have noncircular cross-sections. For example, cotton fiber has a cross-section that appears like a collapsed circle. Wool is more or less circular in cross-section but like most animal fibers has scales on the surface. Silk fiber has a triangular cross-section that is responsible for its luster and the silky feel. The cross-sectional shape of a synthetic fiber can be varied by changing the shape of the spinneret holes through which the polymer is extruded. In fact, one can even produce hollow fibers. Synthetic fibers can have a dog bone, kidney bean, or trilobal cross-sectional shapes. A trilobal cross-section of fibers is useful in promoting bulkiness and cover in pile fabrics such as carpets. Most melt spun fibers, however, have a round cross-section. Solution spinning can result in lobed or dog bone type cross-sections even though the spinneret orifice may be circular. The reason for this is that although the jet emerging from the orifice is circular, it collapses as the solvent is removed by drying (in dry-spinning) or by solution (in wet spinning). Such noncircular fibers can provide a greater density in a fabric than circular fibers. Figure 2.13 shows schematically the variety of cross-sectional shapes that is commonly obtained in fibers.

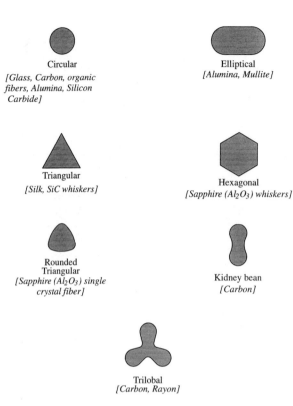

Figure 2.13 Fibers can have variety of cross-sectional shapes, although most fibers have a more or less circular cross-section.

Circular
[Glass, Carbon, organic fibers, Alumina, Silicon Carbide]

Elliptical
[Alumina, Mullite]

Triangular
[Silk, SiC whiskers]

Hexagonal
[Sapphire (Al$_2$O$_3$) whiskers]

Rounded Triangular
[Sapphire (Al$_2$O$_3$) single crystal fiber]

Kidney bean
[Carbon]

Trilobal
[Carbon, Rayon]

The surface texture of fibers can also vary considerably. Vegetable kingdom fibers such as cotton, jute, sisal, etc., have an irregular surface, while most animal fibers such as hair have scales. Compared to the natural fibers, synthetic fibers generally have relatively smooth surfaces. However, synthetic fibers can also show a large degree of variation in their surface texture. Striation markings due to drawing are commonly observed on the surface of metallic filaments. Figure 2.14 shows an example of such markings on the surface of a tungsten filament. Similar markings but on a finer scale are observed on the surface of carbon fibers. Very fine-grained ceramic fibers can show surface roughness that scales with the grain size of the fiber (Chawla *et al.*, 1993). Such surface roughness characteristics can have important bearing in regard to mechanical bonding of fibers with matrix materials in different composites (Chawla, 1993).

2.2.8   Density

The density of fibers is a very important attribute. We mentioned above the use of linear density or mass per unit length. That term is really suitable for giving an idea of the fiber size. Bulk density or mass per unit volume tells us how heavy a material is. Bulk density values of some important fibers are given in Table 2.2. The reader can easily verify that if we divide the linear density of a fiber by its

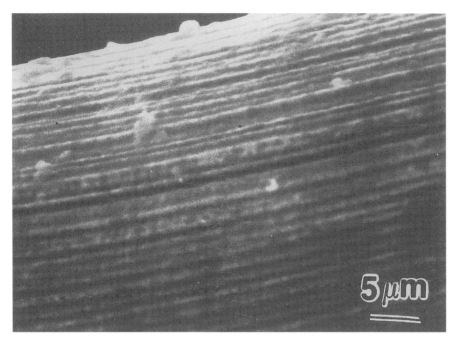

Figure 2.14 Surface markings on a tungsten filament resulting from the die used for wire drawing (SEM).

bulk density, we get the cross-sectional area of the fiber. Most polymeric fibers have bulk densities varying between 0.9 and 1.5 g cm^{-3} (polypropylene 0.9, poly-ethylene 0.97, polyester 1.4, cotton 1.5 g cm^{-3}). Ceramic fibers have densities varying between 2 and 4 g cm^{-3} while metallic fibers can have densities in the range 2 to 20 g cm^{-3}. The closer the value of the density of a fiber to its theoretical value, the smaller will be the amount of microvoids in the fiber and the closer it will be to its theoretical composition. For example, the theoretical bulk density of carbon is 2.1 g cm^{-3}, while the actual bulk density of carbon fiber can range from about 1.6–1.8 g cm^{-3} to close to 2 g cm^{-3} for highly aligned graphitic fibers obtained from a pitch precursor. The reason for this is that the pitch-based graphitic fibers have fewer non-carbon atoms and microvoids than ordinary carbon fibers.

2.3   Important fiber types

One may divide the whole field of fibers in many different ways. One may divide them as natural and synthetic fibers or as polymeric, metallic, and ceramic fibers, etc. One convenient classification is based upon the fiber end use, i.e. apparel and nonapparel fibers. The apparel fibers include synthetic fibers such as nylon, rayon, polyester, spandex, and natural fibers such as wool, cotton, jute, sisal, ramie, silk, etc. Here, it would be in order to take care of some ambiguities regarding nomenclature of some common fibers. The names such as nylon, rayon, spandex are generic names, i.e. these terms represent a group of similar materials much as does steel, or glass or carbon. As such, we spell them *without* a capital letter in this text. Nonapparel fibers include aramid, polyethylene, steel, copper, carbon, glass, silicon carbide, and alumina. These nonapparel fibers are mainly used for structural applications such as fiber reinforcements in a variety of composites. These fibers possess higher modulus and strength and a lower strain to failure than the apparel fibers. They are also characterized by rather difficult processing and drastic strength degradation by the presence of small flaws, i.e. they generally have a low toughness.

Table 2.2   *Properties of some important synthetic fibers.*

Materials (fibers)	Tensile modulus (GPa)	Tensile strength (GPa)	Compressive strength (GPa)	Density (g/cm³)
Steel	200	2.8	—	7.8
Al-alloy	71	0.6	—	2.7
Ti-alloy	106	1.2	—	4.5
Alumina	350–380	1.7	6.9	3.9
Boron	415	3.5	5.9	2.5–2.6
SiC	200	2.8	3.1	2.8
S-glass	90	4.5	>1.1	2.46
Carbon P100 (pitch-based)	725	2.2	0.48	2.15
Carbon M60J (PAN-based)	585	3.8	1.67	1.94
Kevlar 49	125	3.5	0.39–0.48	1.45
Kevlar 149	185	3.4	0.32–0.46	1.47
PBZT	325	4.1	0.26–0.41	1.58
PBZO	360	5.7	0.2–0.4	1.58
Spectra 1000	172	3.0	0.17	1.0
Vectran	65	2.9	—	1.4
Technora	70	3.0	—	1.39
Nylon	6	1.0	0.1	1.14
Textile PET	12	1.2	0.09	1.39

2.4   General applications

There are many applications where fibers can be used in the form of a yarn or fabric (woven, nonwoven or knitted), e.g. optical fibers, ropes, cords, woven or knitted fabrics for clothing, nonwovens in the form of felts etc. for paper, blankets, insulation or filtration purposes, and last but not least, as a reinforcement in composites. In all of these applications, it is easy to see that the attribute of fiber flexibility is very important. As an example, let us examine an application involving fiberoptic technology, flexible fiberscopes. These instruments provide a flexible probe and working length that allow one to obtain images as deep as 6 m inside a machine or structure where straight line access is not possible. Such instruments are also being used in medical diagnostics. Thousands of flexible optical glass fibers make a fiberoptic image bundle that carries a live image of an article to the eyepiece. We discuss optical glass fibers in Chapter 7.

Besides the commonplace uses as textiles and the more sophisticated uses as reinforcements to make composite materials, woven fabrics made of a fibrous yarn can be used to make versatile yet low cost structures. Such construction alternatives can provide the following advantages:

- Fabrics can be translucent and thus result in energy efficiency.
- They can meet diverse needs, e.g. recreational, industrial, and defense requirements.
- Extreme weather conditions, ranging from frozen arctic to hot deserts and strong winds can be handled.
- There is virtually no size limit to air-supported structures made of fabrics.

A variety of structures can be made by fabrics. In the soft shell structure, a tensile loading of the fabric occurs, while in air-supported structures, the fabric is put in tension by a very slight internal pressure. A common example of air-supported structures made of light weight fabrics are the bubbles used for sports activities. Figure 2.15 shows such a bubble on a swimming pool during the winter season at New Mexico Institute of Mining & Technology. The space inside the bubble is heated and the hot air supports the bubble. In tension-membrane structures, the fabric is supported by a structural skeleton of arches, masts or cables. There are many examples worldwide where some fabric made of fibers has been used for structural purposes. Just to give an example of such a structure we cite the Denver airport in the US. Its main terminal building has a 45 000 m², tent-like structural roof that is made of a Teflon-coated glass fiber fabric. Teflon is the trade name of Du Pont for polytetrafluoroethylene (PTFE). The fabric is lightweight, translucent and lets in the natural light, and thus results in a reduction in the construction and building operating costs. The building consists of 17 tent

Figure 2.15 A bubble made of nylon fabric on a swimming pool during the winter season at New Mexico Institute of Mining and Technology.

modules supported by two rows of masts, 30 to 37 m high. The *nonstick* Teflon coating gives protection from chemical attacks and improves the dirt-removing effects of rain.

Paper of course is perhaps the most commonplace example of a fibrous product. Although most common paper products are made of cellulosic fibers, paper-like products can also be made from the so-called high performance fibers such as aramid, glass, carbon, or other ceramic fibers.

Use of fibers as reinforcements to make composites is, of course, well established. These are structural applications where, because of the characteristically long length of fibers, they are incorporated in a continuous medium, called the matrix. We describe some of these applications in subsequent chapters in more detail. Yet another common use of fibers of various kinds is in making ropes. In prehistoric times, ropes were made of braided leather strips and vines. Later, vegetable fibers such as jute, hemp, etc., were used to make ropes. More recently, ropes have been made of synthetic polymers and metallic fibers. Ropes can be made by a variety of construction methods: twisted, braided, plaited, parallel core and fiber, and wire rope.

Yet another example of a unique application where materials in a fibrous form are necessary is that of filters of all kinds. We have already mentioned one commonplace example of a fibrous product, namely a tea bag. The proper tea bag needs to be porous, should not impart a taste to the brew, and of course, should have enough wet strength so that it does not fall apart in the hot water. Filters made of fabrics are commonly used for air cleaning. Industrial baghouses are commonly used to collect dust and particles in various operations where dust laden gases are passed through a filtration medium. The dust (e.g. carbon black, fly ash, etc.) is collected on the filter and the clean gases come out of the other side. Woven or felt types of fabrics are used for filters. Woven fabrics offer lower resistance to gas flow than nonwovens. They also have a smooth finish which provides an easy release of dust. Essentially, the woven structure provides a grid to capture the particulate matter. Nonwoven fabrics provide a greater resistance to gas flow and they work better with heavier dust such as sand, limestone, grain, etc. A measure of how much gas is driven by unit area of filter is given by

Gas volume (G)/cloth area(C)$=G/C$ ratio

In very simple terms, a filter fabric intercepts a particle moving along a gas stream. When a particle impacts a fiber in the filter, the fiber does not get out of the way and the particle cannot go around the fiber. The gas stream, of course, can go around the fiber. The particle has mass and thus cannot follow the gas stream. Very small and light particles can be influenced by the bombardment of gas molecules, changing their path until they bump into the fiber and are trapped. Then there is also electrical entrapment when fibers in the fabric and dust particles carry opposite charges. In this case particles are attracted to the fiber and are trapped there. Fabrics made of cotton, polypropylene, nylon, polyester, Nomex, Teflon, glass, and wool are used as filters.

With the commercial availability of high temperature ceramic fibers, filters made of ceramic fibers such as Nextel 312 (alumina+silica+boron) are used as candle filters.

Fiber characteristics that should be taken into account for making filters include: temperature capability, corrosion resistance (withstand acids, alkalies, solvents, etc.), hydrolysis (humidity levels), dimensional stability, and cost. Among important medical application of fibers, we should mention sutures and filters. Sutures are used to close the wound during surgical operations. Fibers used as sutures can be absorbable or nonabsorbable. The main requirements are that the suture must have only a minimal amount of reaction with the tissue and, in the case of absorption, there must be minimal chemical irritation (Lyman, 1991). Absorbable sutures are collagen sutures, polyglycolic acid and its lactide copolymers, and polydioxanone. Nonabsorbable sutures include silk, cotton, polyethylene, polypropylene, nylon, PET, and stainless steel. These fibers can be used as monofilaments or multifilaments (twisted or braided). Figure 2.16 shows examples of filters for medical purposes made of polyester, nylon, and metallic fibers. In summary, fibers are used in almost everything that we can see in our

(a)

200 µm

Figure 2.16 Examples of filters for medical purposes made of (a) polyester, (b) nylon, and (c) stainless steel.

(b)

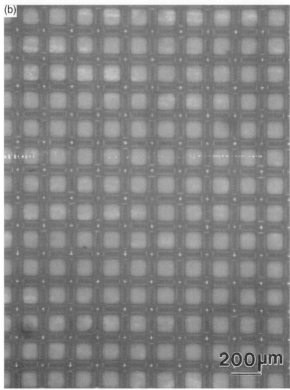

200µm

(c)   **Figure 2.16** (*cont.*)

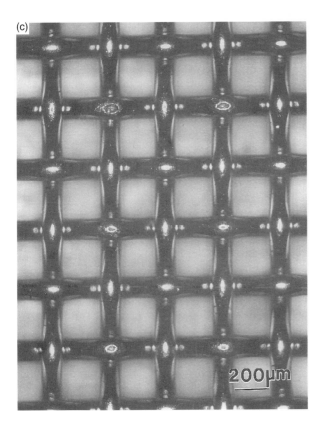

daily life. They range from vegetable to synthetic polymeric fibers to metallic, glassy, and ceramic fibers. The final useful product is, more often than not, based upon a woven, knitted, braided, or a twisted rope or cord form of fiber, although nonwoven fibrous products are also used in less demanding situations, e.g. insulation pads and blankets, etc.

2.5   Health hazards

There are certain potential health hazards associated with the handling of fibrous materials. These came to prominence when the health hazards associated with asbestos fiber came to light (see Chapter 6). We provide a summary of these. The following characteristics of fibers relate to potential biological activity:

- physical dimensions;
- exposure or airborne concentration;
- solubility in the lung.

Physical dimensions or size of a fiber determine whether or not it is respirable. For example, for silica-based glass fiber, fibers less than $3\,\mu m$ in diameter and less than 200 μm in length can become respirable. In terms of exposure or air-borne concentration of fibers, the Occupational Safety and Health Agency (OSHA) of the United States has proposed a permissible exposure limit of 1.0 fiber/cm^3 during any processing or handling. Fiber solubility refers to the speed at which fibers disappear from the lung. In all cases, good work practices keep the exposure to respirable fibers quite low. Gloves and eye protection should be worn at all times.

Chapter 3

Natural polymeric fibers

Commercially, polymeric fibers are perhaps the most important of all, covering as they do a vast range of applications. There are two broad categories of polymeric fibers: natural and synthetic. Natural fibers can be from the vegetable kingdom such as cotton, sisal and jute or from the animal kingdom such as wool, silk, etc. Natural fibers are mostly polymeric in nature. There are, however, some natural fibers that occur in rock formations. These fibers are minerals and can therefore be treated as a ceramic, e.g. asbestos and basalt. We describe these natural mineral fibers in Chapter 6. In this and the following chapter, we describe polymeric fibers. We first briefly review some of the fundamental aspects of polymers and then describe the natural polymeric fibers. We devote Chapter 4 to the synthetic polymeric fibers, which have seen tremendous advancement in the last half of the twentieth century. A vast range of natural polymeric fibers is available and they find large-scale commercial applications. Much research effort is focused on a very special natural fiber originating in the animal kingdom, spider silk. The idea here is to learn about the processing, structure, and properties of silk fibers which are very strong and stiff. More about this later in this chapter. The volume of other natural fibers such cotton, jute, sisal, ramie, etc. in industrial and non-industrial applications has always been quite large because of their many attributes such as the wear-comfort of cotton and the fact that all natural fibers represent a renewable resource. The main disadvantage of natural fibers is the immense variability in their physical, chemical, and mechanical attributes.

First, in order to understand the processing, structure, and properties of polymeric fibers, the main focus of Chapters 3 and 4, it will be useful to review some general and basic concepts regarding the structure of polymeric materials, and especially, some of the important features associated with the macromolecular chains that do not have their counterparts in metals and ceramics. Readers more versed in these aspects of polymers may skip to Section 3.2. In what follows, we first examine the salient structural aspects of polymers and then describe the processing, structure, and properties of some important natural polymeric fibers.

3.1   Structure and properties of polymers

Polymers, natural or synthetic, are characterized by an extended chain structure. Giant molecules called *macromolecules* are formed by joining the chain elements in different ways. Each unit or building block of the chain is called a monomer and a polymer results when we join many monomers to form a chain. We describe below some important classes of polymers, their structure, and some of their important attributes.

3.1.1   Classification of polymers

Polymers can be classified in many ways. An easy way to classify polymers is based on the processing used to make them. There are two main types of polymerization process:

(i) *Condensation polymerization.* In this polymerization, a stepwise polymerization reaction of molecules takes place with a molecule of a simple compound forming at each step. Generally, the simple by-product is a water molecule.

(ii) *Addition polymerization.* In this process, polymerization occurs without the formation of any by-product. This type of polymerization is generally done in the presence of catalysts.

Another convenient classification of polymers involves the behavior of polymer when heated. A *thermoplastic* polymer softens on heating. Many linear polymers show this behavior. Such thermoplastic polymers can be processed by liquid flow forming methods. They also have the advantage of being recyclable. Examples include low and high density polyethylene, polystyrene, polymethyl methacrylate (PMMA), etc. When the molecules in a polymer are cross-linked in the form of a network, they do not soften on heating. These cross-linked polymers are called *thermosetting* polymers. Thermosetting polymers decompose when heated. Examples include epoxy, phenolic, polyester, etc.

We can also classify polymers based on their structure. The basic structural unit of a polymer is a flexible macromolecular chain (Fig. 3.1a). When there is no long-range molecular order and the flexible molecular chains are arranged

(a)

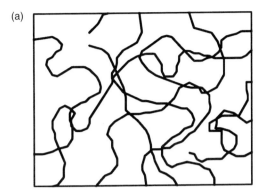

Figure 3.1
(a) Amorphous structure
in a polymer with no
apparent order among the
molecules and the chains
arranged randomly.
(b) Crystalline, lamellar
structure in a polymer
where long molecular
chains are folded in a
regular manner.

(b)

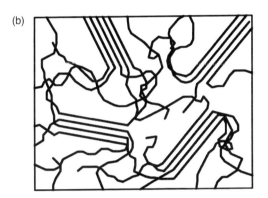

randomly, we have an amorphous structure; such a polymer is called an amorphous or glassy polymer. Under certain conditions, one can obtain a crystalline or, more appropriately, a semi-crystalline polymer. Lamellar crystals form when a crystallizable polymer such as a linear polymer is cooled very slowly from its melting point. Small, plate-like lamellar single crystals can also be obtained by precipitation of the polymer from a dilute solution. In such lamellae, long molecular chains are folded in a regular manner as shown in Fig. 3.1b. Many such lamellar crystallites can group together and form what are called *spherulites*.

Yet another type of classification of polymers is based on the type of repeating unit. A *homopolymer* has one type of repeat unit. *Copolymers* are polymers that have more than one type of monomers or repeat units. If the monomers in a copolymer are distributed randomly along the chain, it is called a *regular* or *random copolymer*. If, on the other hand, a sequence of one type of monomer is followed by a sequence of another type of monomer, it is called a *block copolymer*. If the main chain is one type of monomer and the branch chains are of another monomer, it is called a *graft copolymer*.

Next we describe some of the important attributes and properties of polymers.

3.1.2 Degree of crystallinity

Unlike metals and ceramics, a 100% crystalline polymer is very difficult to obtain. In practice, depending on the polymer type, molecular mass and crystallization temperature, the amount of crystallinity in a semi-crystalline polymer can vary from 30 to 90%. Under special circumstances, such as in a gel-drawn polyethylene fiber, the degree of crystallinity can be as high as 98% while in the case of polydiacetylene, single crystal fibers have been obtained by solid state polymerization (Young, 1989). The inability to attain a fully crystalline structure stems mainly from the long chain structure of polymers. Twisted and entangled segments of chains that become trapped between crystalline regions never undergo the conformational reorganization necessary to achieve a fully crystalline state. Molecular architecture also has an important bearing on polymer crystallization behavior. Linear molecules with small or no side groups crystallize easily. Branched chain molecules with bulky side groups do not crystallize so easily. For example, linear high-density polyethylene can be crystallized to 90% while branched polyethylene can only be crystallized to about 65%. Generally, the stiffness and strength of a polymer increase with the degree of crystallinity.

3.1.3 Molecular mass

Polymers are structurally much more complex than metals and ceramics. One manifestation of this complexity is the parameter molecular weight or, more appropriately, molecular mass. The molecular mass of a polymer is a very important attribute. Any molecule has a molecular mass. The molecular mass of ordinary compounds, however, is not of great importance for most purposes. The importance of this parameter in the case of polymers stems from their long chain structure. Macromolecular chains can encompass a whole range of molecular masses from very low to very high. Even with a simple polymer chain, say polyethylene, one can have a whole range of molecular weights, structural morphologies, and properties. In general, the mechanical characteristics such as strength improve with increasing molecular mass until a plateau is reached.

3.1.4 Melting point and glass transition temperature

A pure, crystalline material has a well defined melting point. The melting point is the temperature at which crystalline order is completely destroyed on heating. Such a phase transition is absent in an amorphous material. However, an amorphous material can be characterized by its glass transition temperature. Figure 3.2 shows specific volume (volume/unit mass) versus temperature curves for an amorphous and a semicrystalline polymer. When a polymer liquid is cooled, it

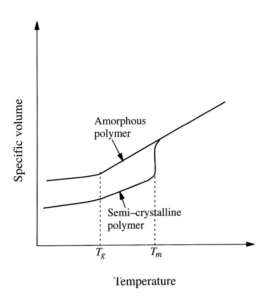

Figure 3.2 Specific volume (volume/unit mass) versus temperature curves for an amorphous and a semicrystalline polymer.

contracts. The contraction occurs due to a decrease in the thermal vibration of molecules and a reduction in the *free space*, i.e. the molecules occupy the space less loosely. In the case of amorphous polymers, this contraction continues to temperatures below the melting point of crystalline polymer, T_m, until the super-cooled liquid polymer becomes rigid due to its extremely high viscosity. Although there is no sharp transition like a melting point, we can define a transition temperature, T_g, corresponding to the onset of rigidity in the material. Note that the structure of an amorphous material below T_g is still that of a liquid. As this is a phenomenon typical of glasses, we call the transition temperature the glass transition temperature. The terms glass and amorphous are used interchangeably. Most inorganic glasses are silica based. Some metallic alloys can be produced in a glassy state by very rapid quenching. In the case of amorphous polymers, we have a glassy structure made of organic molecules. The glass transition temperature, T_g, is in many ways akin to the melting point, T_m, for the crystalline solids. Many physical properties (e.g. viscosity, heat capacity, elastic modulus, expansion coefficient, etc.) change abruptly at T_g. Polystyrene, for example, has a T_g of about 100°C and is therefore rigid at room temperature. Rubber, on the other hand, has a T_g of about −75°C and, therefore, is flexible at room temperature. The glass transition temperature of a polymer depends on its chemical structure. If a polymer has a rigid backbone chain or bulky sidegroups, then its T_g value will be higher than that of a polymer without a rigid backbone or bulky sidegroups. Table 3.1 gives typical T_g values of some common polymers.

Glass transition temperature is a characteristic of any material having a glassy structure, organic polymers, inorganic silica-based glasses, or even metallic glasses. The glass transition temperature, T_g, of inorganic glasses, however, is several hundred degrees Celsius higher than that of organic polymers. The reason for this is the different types of bonding and the amount of cross-linking in organic polymers and silica-based inorganic glasses. Silica-based glasses have mixed covalent and ionic bonding and are highly cross-linked. This gives them a higher thermal stability than polymers which have covalent bonding only and less cross-linking than is found in inorganic glasses. The glass transition temperature of a polymer increases with its molecular mass. This is because the larger the molecule, the less mobile it is, and consequently one needs to raise its temperature higher to obtain sufficient energy for the chains to slide. A higher degree of cross-linking also makes chain sliding more difficult and a higher glass transition temperature will result. On the other hand, the introduction of a light molecule in a high molecular mass polymer will reduce the glass transition temperature by a certain amount. In other words, the range of the molecular mass of a polymer determines the range of its glass transition temperature. This is very important because, more often than not, polymers do show a range of molecular masses and, consequently, they have a range of glass transition temperatures rather than a fixed glass transition temperature.

3.1.5 Mechanical behavior of polymers

Polymeric materials show a wide range of stress–strain characteristics. One characteristic of polymers that is markedly different from metals and ceramics is that their mechanical properties are highly time- and temperature-dependent. An elastomer or a rubbery polymer shows a stress–strain curve that is nonlinear.

Table 3.1 *Glass transition temperatures, T_g, for some common polymers.*

Polymer	Glass transition temperature (°C)
Polyethylene[a]	−80 to −125
Polypropylene[a]	−18 to −25
Polystyrene	100
Nylon 66[a]	50
Poly (methyl methacrylate)	120
Natural rubber	75

Note:
[a] These have a crystalline component, which will also have a melting point.

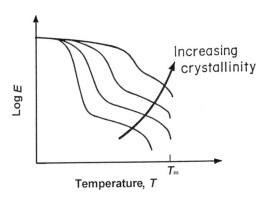

Figure 3.3 Variation of elastic modulus of a polymer with temperature. As the degree of crystallinity increases, the extent of rubbery plateau region decreases.

The characteristically large elastic range shown by elastomers results from an easy reorganization of the tangled chains under the action of an applied stress. Polymers, much more than metals and ceramics, show temperature dependence of their elastic moduli. Figure 3.3 shows schematically the variation of elastic modulus of a polymer with temperature. Also shown is the trend with increasing crystallinity. As the degree of crystallinity increases, the extent of the rubbery plateau region decreases. At a temperature below the glass transition temperature, T_g, the polymer is hard and a typical value of elastic modulus would be about 5 GPa. Above T_g, the modulus value drops significantly and the polymer shows a rubbery behavior.

Yet another important aspect of polymers is associated with environmental effects: polymers can degrade at moderately high temperatures and through moisture absorption. Absorption of moisture from the environment causes swelling in the polymer as well as a reduction in its glass transition temperature, T_g. Polymers can also undergo chain scission when exposed to ultraviolet radiation.

3.2 Natural fibers

Natural polymeric fibers can originate in the animal kingdom or vegetable kingdom. We give below a brief description of these.

3.2.1 Animal kingdom fibers

Animal fibers are basically made up of some kind of protein. Collagen, a crystalline material, is the basic structural fiber of the animal kingdom. It is found in virtually all animals, usually as a component of a complex plant connective tissue. Its structure is based on a helical arrangement of three, noncoaxial, helical polypeptides, stabilized by interchain hydrogen bonds. The three helical polypeptides form a second-order helix with a long pitch around a central axis to give a coiled structure. This is analogous to a three-strand rope, each strand consisting of a polypeptide chain.

All fibers obtained from the animal kingdom consist of elongated protein crystals, aligned more or less along the fiber axis. The protein crystals are embedded in a protein matrix, thus forming a kind of natural composite. Keratin is a general term for fibrous, proteinaceous products of epidermal cell of vertebrates. Human hair is a natural polymeric fiber that we are all familiar with. Figure 2.10 in Chapter 2 shows the scaly surface texture of human hair. Human hair is a mammalian keratin and is very flexible because of its small diameter. Horn, hoof, quill, etc., on the other hand, contain stiff keratins and make rather stiff fibrous composites. Other examples of protein-based fibers include hair-like fibers such as wool and cocoon fibers such as silk. Wool is the fleece obtained from sheep. Wool fibers are generally between 20 and 50 μm in diameter. The main constituent of wool is also keratin. Its strain to failure is about 30%, and like cotton it can absorb moisture, up to one-third of its weight. It can be easily dyed to a variety of colors. Wool also has an outer membrane, called the epicutile, which repels water. All these characteristics make wool a very versatile and prestigious textile fiber. Wool fiber, like many other protein-based natural fibers, shows a scaly texture on the surface. This leads to a directional frictional effect, friction being greater when pulled against the scales than when pulled along the scales (Makinson, 1972). Sliding against scales is difficult because of the roughness. Besides its use in clothing, wool fiber finds uses in blankets, insulating materials, upholstery, carpets and rugs, and felts. Other wool-like fibers are obtained from goat, camel, etc. They are, however, more commonly referred to as hair. The hair of an Angora goat is termed *mohair*.

All natural fibers derived from animals have a proteinaceous constituent: collagen, keratin, or fibroin. Fibroin is the fibrous constituent in silk fiber (see below). The basic building block of a protein is an amino acid which has the following formula:

$$H_2N-\overset{\displaystyle \overset{H}{|}}{\underset{\displaystyle \underset{R}{|}}{C}}-CO_2H$$

where ($-H_2N$) is the amino group, ($-CO_2H$) is the acidic group and R is a side group. The amino and acidic groups react to form an amide linkage ($-CO-NH-$), as in a polyamide (see Chapter 4). The distinctive feature of protein-based fibrous materials is that the R group is different in different amino acids. The following chemical formula shows the linkage of amino acids via an amide linkage ($-CO-NH-$):

$$H_2N-\underset{\underset{R_1}{|}}{\overset{\overset{H}{|}}{C}}-\overset{\overset{O}{\|}}{C}-\underset{\underset{H}{|}}{\overset{\overset{H}{|}}{N}}-\underset{\underset{R_2}{|}}{\overset{\overset{H}{|}}{C}}-\overset{\overset{O}{\|}}{C}-\underset{\underset{H}{|}}{\overset{\overset{H}{|}}{N}}-\underset{\underset{R_3}{|}}{\overset{\overset{H}{|}}{C}}-CO_2H$$

Note the three different side groups, R_1, R_2 and R_3. That is what makes collagen, keratin and fibroin different from one another.

Silk fiber

Silk fiber is said to have originated in China, from where it is thought to have spread to Japan, India, and Iran, and eventually to Europe. Silk fabrics have been highly valued through the ages because of silk fiber's strength, elasticity, softness, affinity for dyes, etc. Fabrics made of silk fiber form part of the folklore all over the world. It has been a luxury fiber for making dresses, Japanese *kimono*, and Indian *sarees*. Although many synthetic fibers have replaced silk, partially or completely, in some products, it still retains a great appeal for luxury apparel and specialized goods. For the reader interested in the allure of silk through the ages, there is ample information available in the lay literature, e.g. Feltwell (1990), Hyde (1984).

Silk fiber is the secretory product of the spinning glands found in a variety of insects and spiders. Different spinning glands of an insect produce different kinds of silk, but all silks are proteinaceous and belong to the fibroins. Although a variety of insects produce silk fiber, only *bombyx mori* and *antheraea pernyi*, two species of moth, are commercially grown for textile purposes. Of these two, the more important variety of silk fiber is the one secreted as a continuous filament by the *bombyx mori* silk worm. The practice of cultivation of the silk moth is called *sericulture*. The cocoon variety of silk has been around for a long time and is a well-established industry. In the sericulture practice, silk worms are fed on leaves from mulberry or oak trees. The silk worm has two sacs of fibroin, a crystalline protein. In the process of silk fiber production, the worm attaches itself to a twig and extrudes the fibroin by a muscular action through a hole or spinneret. Silk fiber (produced by a silk worm or spider) has a complex structure consisting of two strands of fibroin surrounded by an amorphous sheath of sericin, a gummy material. This gummy material is usually removed before spinning the fiber into yarn. The extended protein chains that form the fibroins consist of amino acids, predominantly glycine (44%) and alanine (30%). There is a growing interest in exploring other varieties of silk as well as learning from nature, especially the spider, about the processing and microstructural design of silk fibers. As a spider spins silk fiber, the water-soluble liquid transforms to the insoluble solid silk fiber. The exact mechanism of this transformation is not yet fully understood, but apparently the molecules of the polypeptide chain change

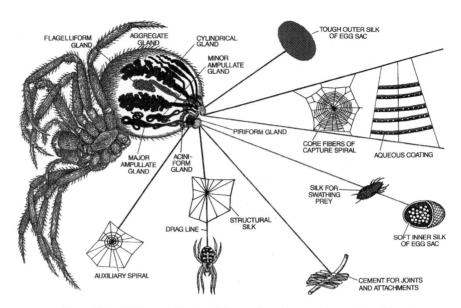

Figure 3.4 Different silks for different functions can be produced by the same spider (after Vollrath, 1992).

their configuration. In a silk worm it is known that this happens simply by pulling on the chain. The reason for the interest in spiders is that they produce a variety of silk fibers that show superb mechanical characteristics, are produced at ambient temperatures and with an aqueous solvent, and are biodegradable. Compare this to the processing conditions required for making Kevlar aramid fiber (see Chapter 4), such as the use of boiling sulfuric acid as a solvent, etc. Hence, the great interest in understanding the structure and properties of spider silk, i.e. the materials science of these spider silk fibers.

All spiders produce silk, but the orb-web spinning *araneid* and *uloborid* spiders are of special interest because they spin seven types of silks. Orbs are 'two-dimensional, point-symmetrical arrays' that form a silk frame enclosing a series of radiating lines (Shear, 1994). There are other types of aerial webs, very different from the orb, but we shall not discuss these. Spider silk is one of the strongest materials known. Two of these seven spider silks, frame silk and viscid silk, have been studied in some detail (Gosline *et al.*, 1986, 1995; Vollrath *et al.*, 1990, Kerkam *et al.*, 1991). The frame silk is the one that is used to make the web frame and radii of the orb-web and the dragline. The viscid silk forms the spiral of the orb-web that is used for catching of the prey. A spider can produce more than 30 m of silk in a single stretch. Different silks for different functions can be produced by the same spider as seen in Fig. 3.4 (Vollrath, 1992). Different silks can have different amino acid compositions. These silks are synthesized and spun by different glands and spinnerets located on the posterior end of the spider's abdomen. Seven kinds of silk are produced from abdominal glands and spigots.

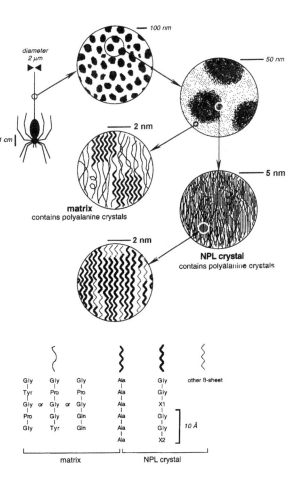

Figure 3.5 Detailed structure of the major ampullate gland silk. All microstructures are in the longitudinal direction. NPL = non-periodic lattice (after Thiel and Viney, 1995).

For example, in the major ampullate gland silk, glutamine (Glu), proline (Pro), glycine (Gly), and alanine (Ala) form 80% of the silk. The amount of Pro can vary considerably from one species to another. Although each gland has its own shape and size, their function is more or less the same. The major ampullate gland silk has been studied the most because of its easy accessibility. One such detailed structure, shown in Fig. 3.5, is the hierarchical structure in the longitudinal direction of major ampullate gland silk (MAS) (Thiel and Viney, 1995). The acronym NPL in the figure stands for a new term coined by these authors, namely 'non-periodic lattice' crystals. These NPL crystals have an ordered molecular arrangement that does show diffraction although it does not contain true long-range periodicity. Figure 3.5 shows short irregular sequences in the NPL crystals while longer irregular sequences occur in the matrix region. Clearly, the structures of silk can become quite complex.

X-ray diffraction studies on frame silk indicate it to be a microcomposite, consisting of a cross-linked, amorphous network (about 55–60%) reinforced by stiff

(a)

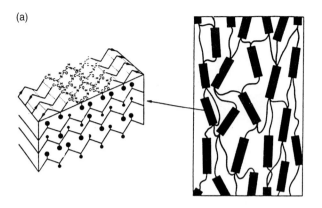

Figure 3.6 Two possible structures of spider frame-silk: (a) microcomposite structure consisting of randomly distributed protein crystallites in an amorphous matrix (after Gosline *et al.*, 1986); and (b) skin–core structure. The two core regions are marked 1 and 2 and the thin skin is marked 3. Both the core regions contain pleated fibril-like strands (after Li *et al.*, 1994).

(b)

1   2   3

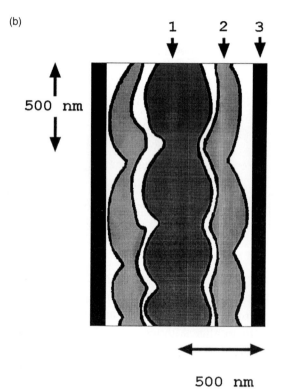

500 nm

500 nm

crystalline protein fibers. Figure 3.6a shows a schematic of the microcomposite structure of the frame silk. Li *et al.* (1994) made a three-dimensional, nanometer scale study of the structure of dragline or frame silk by atomic force microscopy. The three-dimensional structure of silk was determined from longitudinal sections and sections made at 45° and 90° to the fiber axis. The structural model of spider dragline silk, due to Li *et al.* (1994), is shown in Fig. 3.6b. Unlike the structure shown in Fig. 3.6a (randomly distributed protein crystallites in an amorphous matrix) their model consists of an organized skin-core structure. In

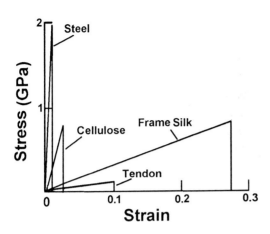

Figure 3.7 Tensile stress–strain curves of a variety of different materials: steel, cellulose, tendon, and frame silk. Note the high fracture energy (area under the stress–strain curve) of the frame silk fiber. The strain-to-failure of frame silk can be as high as 30% (after Gosline *et al.*, 1986).

the figure the two core regions are marked 1 and 2 and the thin skin is marked 3. Both the core regions contain pleated fibril-like strands. This model of spider silk structure is very similar to the structure of Kevlar aramid fiber (see Chapter 4).

The tensile stress–strain curve of frame silk is shown in Fig. 3.7. The curves for a number of different materials, steel, cellulose, tendon, are also included for comparison. Note the high fracture energy (area under the stress–strain curve) of the frame silk fiber. The strain to failure of frame silk can be as high as 30%.

However, unlike the cocoon silk fiber from the larvae, the spider silk, although much stronger than the silkworm silk, is not available in large quantities (Vollrath, 1992; Gosline *et al.*, 1986). It is so difficult to produce that a commercial production is not viable. One of the reasons for this is that the spiders need a steady stock of live insects to eat, and if they are put too close to other spiders, say in a spider silk producing factory, they end up eating each other (Berenbaum, 1995). Efforts are being made in the area of genetic engineering to produce silk-like proteins in microbial culture and spin fibers from them (Gosline *et al.*, 1995; Kaplan *et al.*, 1994). Researchers at the US Army Research and Development and Engineering Center (Natick, MA) are developing *E. Coli* bacteria that are genetically altered to produce silk (Kaplan *et al.*, 1994).

The cocoon silk of the silk worm is an excellent textile material and the commercial activity in this realm attests to that. Its mechanical properties, however, are rather modest. This has been attributed to the fact that superior mechanical characteristics are not required in the cocoon. On the other hand, in spider's orb-web silk, the spider needs a fiber that can absorb the impact of the falling insect prey. Hence, the need for superior mechanical characteristics.

Silk is combustible, but self-extinguishing. It is a good electrical insulator and can accumulate static electricity easily, which is the reason why silk fiber processing is done at high humidity. It is 1–3 μm in diameter, has a density of 1.25 g cm^{-3}, a strain to failure of between 20% and 30%, and a tenacity between 3 and 5 gram/denier. The elastic modulus of spider silk is about 10 GPa. The cross-section of a silk fiber is an irregularly shaped triangle, or crescent, or an elongated wedge. It is this more or less triangular cross – sectional shape of a large number of fine silk fibers in a fabric that makes them reflect light like prisms and gives the fabric that characteristic luster.

3.2.2   Vegetable fibers

The vegetable kingdom is, in fact, the largest source of fibrous materials. The great advantage in this case, of course, is that of low cost. Cellulosic fibers in the form of cotton, flax, jute, hemp, sisal, ramie, etc. have been used in the textile industry while wood and straw have been used in the paper industry. We describe now, very briefly, some important vegetable fibers. The interested reader is directed to a number of good review articles for greater details (Batra, 1985; Meredith, 1970; Robinson, 1985; Stout, 1985; Segal and Wakelyn, 1985; Rohatgi *et al.*, 1992, Chand and Rohatgi, 1994).

Fibers having origin in the vegetable kingdom can be obtained from various parts of the parent plant. Some examples of vegetable fibers obtained from different parts of the plant as follows:

- seed hairs (cotton, kapok);
- stem or bast fibers (flax, hemp, jute);
- leaf fibers (piña, sisal).

These fibers serve a variety of functions in their parent plants. Most commonly, they serve to stiffen the stems and leaves of plants. This allows the plants to weather the storms and high winds to which they are frequently subjected. In some plants, they also provide buoyancy to the seeds.

The main constituent of any vegetable fiber is *cellulose*. Cellulose is often found as a relatively high modulus, fibrillar component of many naturally occurring composites, such as wood, where it is found in association with lignin. Cotton fiber, for example, contains 80–85% cellulose, while other vegetable fibers such as ramie, hemp, jute, flax and sisal have between 65 and 70% cellulose. Cellulose is composed of three elements, C, H and O, with the general formula of $C_6H_{10}O_5$ and is crystalline. An easy way to check this is to use an X-ray diffraction technique (Barkakaty, 1976). An amorphous polymer will give diffuse halos. A diffuse halo indicates an irregular atomic arrangement, i.e. an amorphous structure. There may also occur some alignment of chains in the amorphous regions. An X-ray diffraction pattern of a highly crystalline polymer would show regular spots and/or regular rings indicating regular spacing characteristic of an orderly arrangement in a single crystal. Sometimes the discrete spots take the form of streaks or arcs because of a preferential alignment of polymeric chains. Well-spaced regular rings would indicate a random, polycrystalline region. Figure 3.8 shows a Debye–Scherer X-ray diffraction pattern of a sisal fiber. The

Figure 3.8
Debye–Scherer X-ray
diffraction pattern of a
sisal fiber. The arcs in this
pattern indicate the crystalline nature of the main
constituent, cellulose.

arcs in this pattern indicate the crystalline nature of cellulose, the main constituent of sisal. Besides cellulose, vegetable fibers contain significant amounts of
other chemicals such as lignin, hemicellulose, pectin substances, resins, mineral
matter, fats, and waxes. After their isolation from the plant, these fibers are subjected to some treatments to remove hemicellulose and lignin, etc. Most of the

lignin and other non-cellulosic substances are associated with the cell walls and their presence *does* modify the final properties of the fiber. The non-cellulosic material is hardly ever removed completely from these fibers, mainly because it is prohibitively expensive to do so.

An important attribute of vegetable fibers is that they can absorb moisture from the atmosphere in comparatively large quantities. This is because the cellulose is hygroscopic, i.e. it absorbs moisture from the atmosphere at high humidities and releases it when dry. This absorption leads to alterations in weight and dimensions as well as in strength and stiffness. The moisture content can be measured as the percentage loss in weight of the fiber when it is dried in an oven at a given temperature and for a specified time. Frequently, we express the moisture content as 'moisture regain', which is the loss in weight on drying expressed as a percentage of the weight of the dried fiber, i.e. we take the weight of the dry fiber as the basis because the dry weight of the fiber is a constant, whereas the original weight can vary according to the moisture content before drying.

Another important attribute of vegetable fibers is that they are subject to biological decay. Most vegetable fibers darken and weaken with age and exposure to light. They are not as durable as synthetic polymeric fibers. All of them get easily attacked by a variety of organisms, at high humidity and temperature, leading to rot and mildew.

We provide below a brief description of some of the important vegetable fibers.

Cotton

Cotton has been called the king of fibers. Cotton is a hollow fiber when it is growing, but its lumens, the internal space of a cell, collapse when it is harvested. Thus, its cross section is flat and one can see the microfibrils composing the cotton fiber in a low magnification optical microscope. The composition of cotton is mainly crystalline cellulose with varying amounts of pectin, fat and wax. The cellulose molecule takes a helical form with the helix angle varying between 20 and 30°. Figure 3.9 shows the structure of cotton fiber (Jefferies *et al.*, 1969). The density of cotton fiber is about $1.5\,g\,cm^{-3}$.

Cotton is really a very versatile fiber in that it takes color easily, absorbs moisture readily, and is durable in a wide variety of environments. Cotton yarn finds use in all kinds of clothing, often mixed with synthetic fibers. In fact, according to one estimate almost half of all textiles worldwide are made of cotton (Thompson, 1994). Besides apparel, cotton fiber is used in coffee filters, fishnets, lace, upholstery, etc. It is biodegradable and can be used as a sorbent in oil spills.

Jute

Jute fiber is obtained from the inner bark of plants of the genus *corchorus*. It is produced mainly in Bangladesh, Brazil and India. It is used extensively in

Lumen

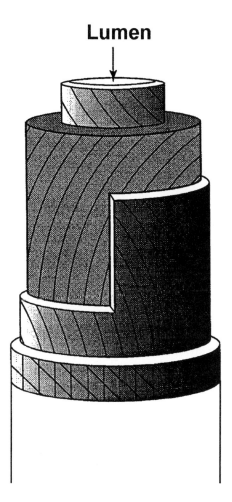

Figure 3.9 Structure of a cotton fiber. The cellulose molecule takes a helical form with the helix angle varying between 20 and 30° (after Jefferies *et al.*, 1969).

making cords, coarse cloth and sacks. Occasionally, one can even find a dinner jacket made of lustrous golden colored jute fiber.

Jute fibers are separated from the woody part of the stem by a process called *retting.* The process involves laying bundles of stems flat in water, side by side, to form a platform. The platform is then covered first with weeds and then with heavy logs to keep the bundles submerged in the water. The period of retting varies with the water type and temperature. The fiber strands are then removed, manually, from the stems. They are cleaned by spreading on the water surface, squeezed to remove the excess water and hung on frames to dry in the sun for 2–3 days. The dried jute fibers are then tied in bundles and packed for shipment.

The jute fiber can have a color from lustrous yellow to brown The commercially obtained strands consist of individual jute fiber cemented together by natural gums. When observed in an optical microscope, the cross section of a jute

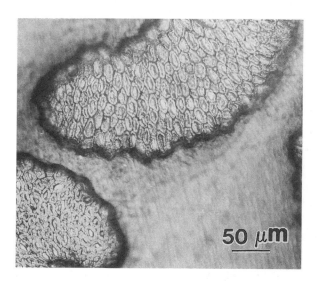

Figure 3.10 Polygonal cells of jute fiber. Note the thick walls and easily observable lumens.

fiber shows polygonal cells. These cells are about 2.5 mm long. The cells have thick walls and easily observable lumens (see Fig. 3.10).

As with most fibers derived from the vegetable kingdom, the individual fibers vary considerably in strength. Their strain to fracture is about 1.7%. The density of jute is 1.5 g cm^{-3}. Jute fibers absorb a considerable amount of water, typically showing a regain value of about 15 wt% at 65% relative humidity.

Sisal

Sisal is a leafy fiber that comes from the plant *agave sisalana* which is native to Central America. Tropical climate with a moderate humidity is required for sisal fiber and it is commonly found in Eastern Africa and Brazil.

The plant has a short, thick stem from which grows a rosette of leaves containing sisal fibers. The leaves are cut off at the point where they join the main stem and are sent for extraction of the fiber. This must be done as soon as possible after harvesting as the cut leaves deteriorate very quickly. The fibers are extracted by crushing or scraping to remove green matter surrounding them. The extracted fiber is dried in the sun. Strands of commercially obtained sisal fiber can vary in length from 0.5 to 1.25 m. Like jute fiber, a large number of individual fibers are generally cemented together by natural gums. The color of sisal fiber varies from white to pale yellow.

Sisal fiber finds use in a variety of items such as in the manufacture of binder twine and baler twine, mattings, rugs, marine ropes, padding material for cars, upholstery, etc.

Ramie

Ramie is another important vegetable fiber. It has a shiny, white color. Its strain to failure is 3–4%, and compared to other vegetable fibers, it has an exceptionally high resistance to bacteria, fungi, and mildew. Often, it is blended with other fibers for use in making clothing.

Rayon

Rayon fiber, produced by conversion of cellulose, is commonly referred to as regenerated cellulose. In these fibers, the primary component is cellulose, originating in the plant kingdom. That is why rayon is generally included in the category of natural fibers. Cellulose from a variety of sources can be used to produce rayon fibers. For example, cellulose pulp obtained from cotton linters (short fibers sticking to the cottonseed) or from timber containing more than 90% cellulose is suitable for making rayon fiber. Cellulose does not melt or dissolve in common solvents. Therefore, it is converted into a different chemical form. For example, to make rayon, one dissolves wood pulp in a solution of sodium hydroxide. Carbon disulfide is added to this to produce sodium cellulose xanthate. This chemical form of cellulose is soluble in dilute, aqueous sodium hydroxide, resulting in a solution with molasses-like consistency, called viscose. Viscose is wet spun through a spinneret into a sulfuric acid bath where the extruded fibers are reconverted to cellulose and coagulated to form solid filaments. The rayon fibers are then wound as continuous fiber on to spools or cut into short lengths to make a staple rayon fiber and washed to remove the processing chemicals. Variations in cross-sectional shape, stretching, dyeing, introduction of crimp, or chopping into short lengths can be done during the spinning process.

A summary of important properties of some vegetable fibers is given in Table 3.2 (Rohatgi *et al.*, 1992).

Table 3.2 *Properties of some vegetable fibers (adapted from Rohatgi et al., 1992).*

Fiber	Density $(g\,cm^{-3})$	Strength (GPa)	Young's modulus (GPa)	Specific strength[a]	Specific modulus[a]
Jute	1.50	0.85	64	0.57	43
Ramie	1.50	0.93	59	0.62	39
Hemp	1.50	0.90	69	0.60	46
Flax	1.50	1.08	100	0.71	67

Note:
[a] Specific strength and modulus are strength/density and modulus/ density, respectively.

3.2.3   Environmental effects on polymeric fibers

Environmental factors such as humidity, temperature, pH, ultraviolet radiation, and micro-organisms can affect polymer fibers. Natural polymeric fibers are more susceptible to environmental degradation than synthetic polymeric fibers. Larvae of some insects may feed on a wide variety of dry materials with a high cellulose and protein content, i.e. vegetable as well as animal fibers. Therefore fibers such as silk, wool, cotton, jute, etc. and the products made of such fibers, e.g. textiles (sweaters, coats, etc.) and carpets are susceptible to damage by micro-organisms. Cellulose, in particular, is attacked by a variety of bacteria, fungi, and algae. Micro-organisms use cellulose as a food source. Such attack can be prevented to some extent by giving a durable moth-resist treatment during manufacture. A good example of such a treatment is that given to wool fiber. Perhaps, the most widely used treating agent is permethrin, a synthetic pyrethroid. Pyrethroid is chemically similar to pyrethrum, an insecticide occurring naturally in a number of chrysanthemum varieties. Permethrin is widely used in agriculture, for the domestic control of insects and as a wood preservative.

Natural fibers based on cellulose or protein are more prone to degradation due to humidity and temperature than synthetic organic fibers. Photo-degradation occurs when exposed to light (both visible as well as ultraviolet) which shows up as a discoloration and loss in mechanical properties. Most polymeric fibers (natural or synthetic) swell due to moisture absorption. The swelling is generally more in natural fibers. We discuss this topic again in Chapter 4.

3.3   Applications of natural polymeric fibers

Natural polymeric fibers find extensive applications in all walks of life. We give some examples.

Natural polymeric fibers are used for a variety of textile applications ranging from clothing to upholstery to sacks and cordage. Frequently, these are blended with synthetic fibers to obtain an optimum set of properties. Many naturally occurring fibers can be and are used in composites, but mostly in applications involving not very high stresses (Chawla, 1976; Chawla and Bastos, 1979; Roe and Ansell, 1985; Chand and Rohatgi, 1994)

Many natural fibers are finding uses in automobiles, trucks, and railway cars. One of the reasons for this is the concern for the environment and the need to use renewable resources. Natural fibers are a renewable resource and they do not exacerbate the CO_2 emission problem. Not only are natural fibers environmentally friendly, they possess reasonably good mechanical properties, such as impact resistance, at a very low weight. Examples of applications of natural fibers include, among others, the Mercedes–Benz E-class automobile. More specifically, in this automobile shredded cotton is used in rear shelves and insulating mats, while padding elements of coconut fiber, animal hair, and natural latex, and structural parts of flax and sisal fiber reinforced polymers are used. It should be pointed out that flax fiber is native to Germany while sisal is native to Brazil and many other countries. Natural fibers are also being used as geotextiles for soil stabilization (Datye and Gore, 1994; Kaniraj and Rao, 1994).

Chapter 4

Synthetic polymeric fibers

In this chapter we describe synthetic polymeric fibers, which saw tremendous advancement in the last half of the twentieth century. In fact, a reasonable case can be made that the so-called *age of fibers* began with the advent of synthetic fibers such as nylon, polyester, etc. in the late 1930s and early 1940s. Many companies such as Du Pont, Monsanto, BASF, Hoechst, ICI, etc. contributed significantly in this area. For a historical account of the scientific and technological progress made in this area, the reader is referred to a study of research and development activities at Du Pont during the period 1902–1980 (Hounshell and Smith, 1988). Most of these synthetic polymeric fibers such as polyester, nylon, etc. have very uniform and reproducible properties. They, however, have a rather low elastic modulus, which restricts them mostly to the apparel or textile market. It was the research work aimed at making *strong* and *stiff* synthetic polymeric fibers for use as reinforcements in polymers, which started sometime in the late 1950s and early 1960s, that resulted in the commercial availability of strong and stiff fibers such as aramid and extended-chain polyethylene. We describe below the processing, structure, and properties of some important synthetic polymeric fibers in some detail.

4.1   Brief history of organic fibers

A brief historical review of the work in the area of organic fibers will be helpful in placing things in perspective. We begin with the discovery of nylon. Nylon was discovered and commercialized by Du Pont in 1938 (Magat and Morrison,

1976). Wallace Carothers of Du Pont is generally regarded as the father of nylon. Nylon first penetrated the silk hosiery market just before the start of the Second World War. In fact, 1988 marked the 50th anniversary of the introduction of nylon silk stockings. Nylon is made by melt spinning. It is a very flexible, knittable, and durable fiber; all these attributes have made it one of the most important fibers for the textile industry. Nylon's high strength, good impact and fatigue resistance also led to its use in the tire industry. It should be noted that the term nylon is a generic term that represents a group of similar materials, in the same vein as glass, steel or carbon. Accordingly, we spell it without a capital letter.

After nylon, poly (ethylene terephthalate) (PET) and polyacrylonitrile (PAN) were the next two major discoveries. PET is made by polymer melt spinning while PAN uses a spinnable polymer solution dope. PET, a thermoplastic polyester, and PAN soon found uses as substitutes for wool, cotton and rayon in carpets, blankets, etc. PAN is also used as a precursor for making carbon fibers (see Chapter 8). Among the major attributes that made the synthetic fibers such as nylon, PET, and PAN, etc. a great success story were the characteristics of wash-and-wear, resistance to wrinkling, and durability. Synthetic polymeric fibers saw tremendous advances in the last half of the twentieth century. More importantly, however, a new era in the age of man-made fibers was launched when the importance of structure–property relationships in polymers was realized. Some impressive work on polymer crystallization in the late 1960s started this new era of structure–property relations in polymeric materials. One of the major outcomes of this structure–property work on polymers was the realization that an *extended and oriented polymer chain* arrangement leads to improved mechanical characteristics. Organic fibers of high tensile strength and modulus, such as aramid and polyethylene, were a direct result of this effort. As we shall see in Section 4.5, two different structural approaches have been used to obtain high stiffness, high strength fibers. In rigid rod polymers such as aramids, we make use of the inherent high stiffness of the aromatic groups in the main chain coupled with the physical ordering of chains, while in flexible chain polymers such as polyethylene, we use the physical ordering of chains to obtain a high modulus.

In what follows, we first describe general processing techniques used to make synthetic polymeric fibers, followed by a description of the processing, structure, and properties of some important low modulus organic fibers. Finally, we describe, in some detail, two commercially important, high-stiffness fibers: aramid and polyethylene.

4.2   Processing

Generally, synthetic polymers are made into fibers by a process known as fiber spinning. Fiber spinning is the process of extruding a liquid through small holes

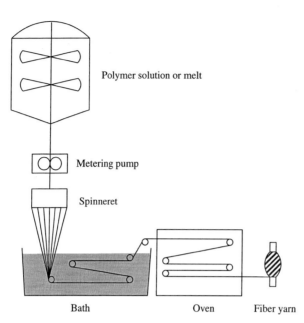

Polymer solution or melt

Metering pump

Spinneret

Bath   Oven   Fiber yarn

Figure 4.1   General view of solution or melt spinning. Filaments emerge out of the holes in a spinneret, pass through a bath and an oven before being wound as a yarn on a bobbin.

in a spinneret to form solid filaments. Figure 4.1 shows a schematic of the process. A polymer melt or solution is the starting material. Filaments emerge out of the holes in a spinneret, may pass through a bath and an oven before being wound as a yarn on a bobbin. It is an unfortunate fact that the term spinning is also used to denote an important process called yarn spinning. In nature, silk worm and spiders produce continuous filaments by this general process. The process of spinning solid fibers by making a liquid pass through tiny orifices exploits the fact that certain materials such as organic polymers and silica-based inorganic glasses (see Chapter 7) have the right viscosity to produce a stable jet as the liquid comes out of the orifice. The jet may be stabilized by an air stream or a coagulating bath (see below). If the viscosity is low, then the molten jet becomes unstable with respect to surface tension because of a surface wave phenomenon. These waves, called Rayleigh waves, form on the surface of the low-viscosity jet stream (see also Chapter 6). Rayleigh waves grow exponentially in amplitude along the length of the jet and thus tend to break the jet up into droplets. Molten metals have very low viscosity (close to that of water), and thus we do not normally use melt spinning to produce metal filaments. The key point to recognize is that if we can somehow stabilize the low-viscosity molten jet against breakup by the Rayleigh waves, then we can have a continuous fiber. One way to do this is to extrude such a low viscosity jet through an orifice into a chemically reactive environment. In the case of glasses and organic polymers, the melts have high enough viscosity (more than 10^4 Pa s or 10^5 poise), which delays the Rayleigh breakup until the molten jet freezes. In the case of a low-viscosity melt, one can avoid the breakup of the molten jet by chemically stabilizing it. For

(a)

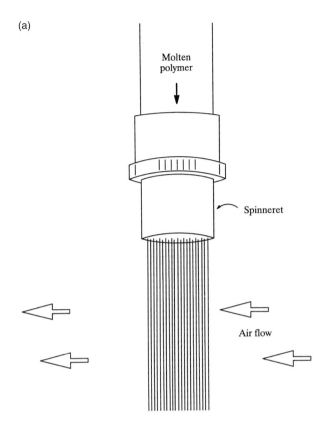

Molten
polymer

Spinneret

Air flow

Figure 4.2 Different
spinning processes. (a)
Melt spinning: the fiber-
forming material is heated
above its melting point
and the molten material is
extruded through a spin-
neret.

example, Wallenberger *et al.* (1992) developed a process to make fibers of
(alumina+calcia), wherein a low-viscosity (1 Pa s or 10 poise) jet is chemically
stabilized with propane before the Rayleigh waves can break the stream into
droplets. The low-viscosity jet must be stabilized in about 10^{-3} seconds or it will
break up into droplets. We discuss this topic further in Chapter 7. Suffice to say
here that in the case of polymers, the viscosity is high enough that it does not
pose the problem of Rayleigh instability leading to a breakup of the molten jet
stream. A brief description of some of the common methods of spinning synthetic polymeric fibers follows:

(a) *Melt spinning.* The fiber forming material is heated above its melting point
and the molten material is extruded through a spinneret (see Fig. 4.2a). The
liquid jets solidify into filaments in air on emerging from the spinneret holes.
Melt spinning is very commonly used to make organic fibers such as nylon, polyester, and polypropylene fibers.

(b) *Dry spinning.* A solution of a fiber forming polymeric material in a volatile
organic solvent is extruded through a spinneret into a hot environment. A stream
of hot air impinges on the jets of solution emerging from the spinneret, evaporates the solvent, and leaves the solid filaments (Fig. 4.2b). Fibers such as acetate,

(b)   **DRY SPINNING**

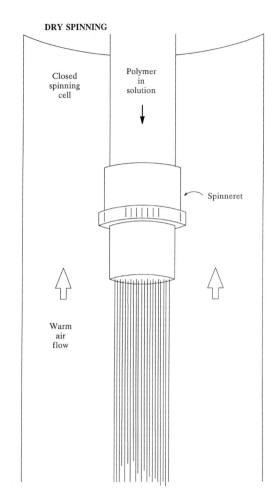

Figure 4.2 (*cont.*)   (b) Dry spinning: a stream of hot air impinges on the jets of solution emerging from the spinneret, evaporates the solvent, and leaves the solid filaments.

acrylic, and polyurethane elastomer are obtained by dry spinning of appropriate solutions in hot air.

(c) *Wet spinning.* A polymer solution in an organic or inorganic liquid is extruded through holes in a spinneret into a coagulating bath. The jets of liquid coalesce in the coagulating bath as result of chemical or physical changes and are drawn out as a fiber. Figure 4.2c shows this process. Examples of organic fibers obtained by this process include rayon and acrylic fibers.

(d) *Dry jet–wet spinning.* Aramid fibers are processed by the dry jet–wet spinning process. In this process, the anisotropic solution is extruded through the spinneret holes into an air gap (about 1 cm) and then into a coagulating bath. The coagulated fibers are washed, neutralized and dried. We discuss this process in detail in Section 4.5.

(e) *Gel spinning.* In this process, filaments coming out of the spinneret holes form a rubbery or gel-like solid on cooling. The process is applicable to linear flexible chain polymers such as polyethylene. A gelled fiber is produced when an appropriate solution coming out of the spinneret is quenched by air. The

(c)

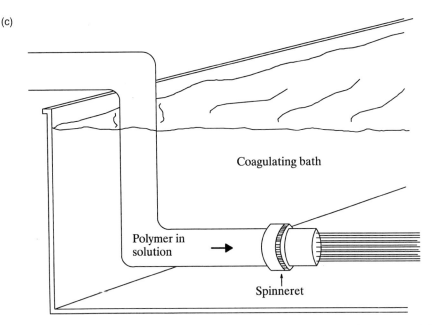

Coagulating bath

Polymer in
solution →

Spinneret

Figure 4.2 (*cont.*)   (c) Wet spinning: the jets of liquid coalesce in the coagulating
bath as result of chemical or physical changes and are drawn out as a fiber.

gelled fiber has the structure of a swollen network of entangled chains, which
allows drawing into fibers with very high draw ratios and below the melting
point. High strength and high stiffness ultrahigh molecular weight polyethylene (UHMWPE) fibers are made by this process as described in Section
4.5.

(f) *Fibers from films*. Techniques (a) through (e), described above, are the
common variants of making polymeric fibers, namely extrusion of a spinning
dope or melt through a large number of small holes in a spinneret. There is a very
different method of making fibrous polymers that involves using films (Krassig
et al., 1984). This technique exploits the ease of making polymer films that are
highly oriented and inexpensive to produce. Such films show a marked
anisotropy in strength along the two principal directions (longitudinal and
transverse), the strength along the longitudinal or draw direction being much
higher than that in the transverse direction. This stems from the preferential
polymer chain alignment in the draw direction. This anisotropy allows easy split-
ting of such films into fibrous products by simple mechanical actions, such as
brushing, rubbing or twisting. The cross-section of such fibers from films is
square or rectangular rather than circular because of the starting material.

4.2.1   Stretching and orientation

We have repeatedly pointed out that a very important characteristic of any polymeric fiber is the degree of molecular chain orientation along the fiber axis. The

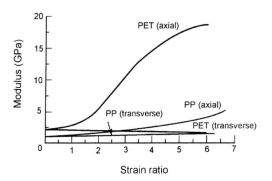

Figure 4.3 Variation of modulus as a function of strain or draw ratio for polyethylene terephthalate (PET) and polypropylene (PP) in the axial and transverse directions (Hadley *et al.*, 1969). Note the rather small effect in the transverse direction.

process of extrusion through a spinneret results in some chain orientation in the filament. Generally, the molecules in the surface region undergo more orientation than the ones in the interior because the near surface molecules are more affected by the edges of the spinneret hole. This is known as the skin effect or skin–core effect (see Chapter 1). It can affect many other properties of the fiber, for example, the adhesion with a polymeric matrix or the ability to be dyed.

Most as-spun polymeric fibers are subjected to some stretching, causing further chain orientation along the fiber axis and consequently better stiffness and strength in tension along the fiber axis. The amount of stretch imparted is generally given in terms of draw or strain ratio, λ, which is the ratio of the initial diameter to the final diameter. For example, nylon fibers are typically subjected to a draw ratio of 5 after spinning. A higher draw ratio results in an increased alignment of chains, and consequently a higher elastic modulus. Figure 4.3 shows the variation of modulus as a function of strain or draw ratio for polyethylene terephthalate (PET) and polypropylene (PP) in the axial and transverse directions (Hadley *et al.*, 1969). Note the rather small effect in the transverse direction.

Orientation of macromolecular chains also affects the ability of a fiber to absorb moisture. An increased chain alignment results in a higher degree of crystallinity and a lower moisture absorption. In general, the stretching treatment translates into a higher resistance to penetration by foreign molecules, i.e. a greater chemical stability. In particular, it affects the dyeing characteristics of fiber, because the dye molecules cannot easily penetrate the fiber. There is, however, a limit to the amount of stretch that can be given to a polymer because the phenomenon of *necking* can intervene and cause rupture of the fiber. Necking results in an accentuated but very localized plastic deformation at a particular site, which eventually leads to fracture of the material at that site. It should be pointed out that the phenomenon of necking can be observed in any material that shows plasticity under tension. An important result of this chain alignment along the fiber axis is the marked anisotropy in the characteristics of a polymeric fiber.

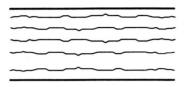

Oriented, long chain polymeric fiber

Figure 4.4 Schematic of anisotropic swelling in a polymeric fiber.

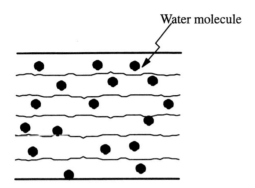

Polymeric fiber + moisture

4.3   Environmental effects on polymeric fibers

Environmental factors such as humidity, temperature, pH, ultraviolet radiation, and micro-organisms can affect polymer fibers. Natural polymeric fibers are more susceptible to environmental degradation than synthetic polymeric fibers. Cellulose is attacked by a variety of bacteria, fungi, and algae. Micro-organisms use cellulose as a food source. Natural fibers based on protein such as wool, hair, silk, etc. can also be a food source for micro-organisms, but such fibers are more prone to degradation due to humidity and temperature. Polymeric fibers, natural or synthetic, undergo photo-degradation when exposed to light (both visible and ultraviolet). Physically this results in discoloration, but is also accompanied by a worsening of mechanical characteristics.

Most polymeric fibers (natural or synthetic) swell due to moisture absorption. The degree of swelling is generally more in natural fibers. In the case of synthetic fibers, especially the ones showing a high degree of crystallinity, and thus anisotropy, more swelling occurs in the transverse direction than in the longitudinal direction. This is because water molecules that penetrate a fiber end up separating polymeric chains laterally more than along the fiber axis. Figure 4.4

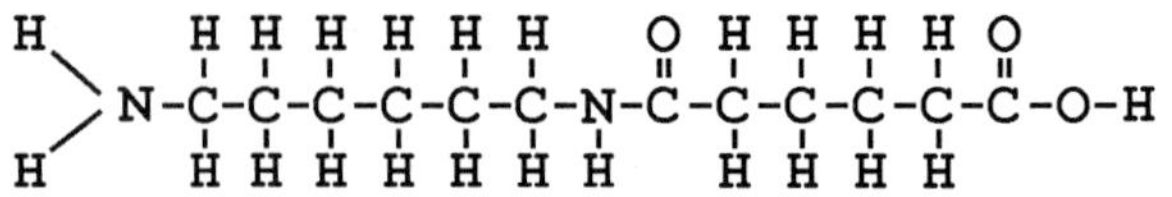

H H H H H H O H H H H O
\ | | | | | | || | | | | ||
>N-C-C-C-C-C-C-N-C-C-C-C-C-C-O-H
/ | | | | | | | | | |
H H H H H H H H H H

Figure 4.5 Chemical structure of nylon.

shows, in a schematic manner, this phenomenon of anisotropic swelling in a polymeric fiber. An important characteristic of woven fabrics is that they have distinct pores (see Chapter 2). Such pores in woven fabrics can close, partially or completely, when fibers swell due to moisture absorption.

4.4   Some important low modulus synthetic polymeric fibers

In this section we describe the processing, structure, and properties of some important low modulus synthetic polymeric fibers.

4.4.1   Nylon

Nylon is a generic name for any long-chain polyamide thermoplastic containing more than 85% aliphatic groups in the main chain. It is worth pointing out that nylon and polyamide are accepted generic names for the same group of fibers. The term nylon is commonly used in North America and UK while polyamide is more prevalent in continental Europe. The chemical structure of nylon 66 is given in Fig. 4.5. The recurring part of the chain, –CONH–, is the amide group. The amide groups in adjacent chains lead to interchain hydrogen bonding and allow close packing. We can use the definition of nylon fiber provided by the US Federal Trade Commission (FTC). According to the FTC, a nylon fiber is a manufactured fiber in which the fiber-forming substance is a long-chain synthetic polyamide in which *less than 85%* of the amide linkages are attached directly to two aromatic rings. This term 'less than 85%' serves to distinguish nylon from aramid fibers (Section 4.5). The term nylon, as we said above, refers to the whole family called polyamides. There are, however, two important types of nylon, nylon 66 and nylon 6, with nylon 66 being the major polyacid fiber. The explanation for this nomenclature of nylon is as follows. Processing of nylon fiber involves making the basic polymer, followed by melt spinning of the fiber. The four basic elements, C, H, N and O are combined into adipic acid hexamethylene diamine and caprolactam. Condensation polymerization provides the long-chain molecule of polyamide. The *66* in nylon 66 refers to the number of carbon atoms in the constituents of nylon, i.e. both diamine and dibasic acid contain 6 carbon atoms each. If, however, the polyamide is obtained by self-condensation of a single constituent (e.g. an amino acid), a single number is used to indicate the number of carbon atoms in the original constituent, i.e. nylon 6. The poly-

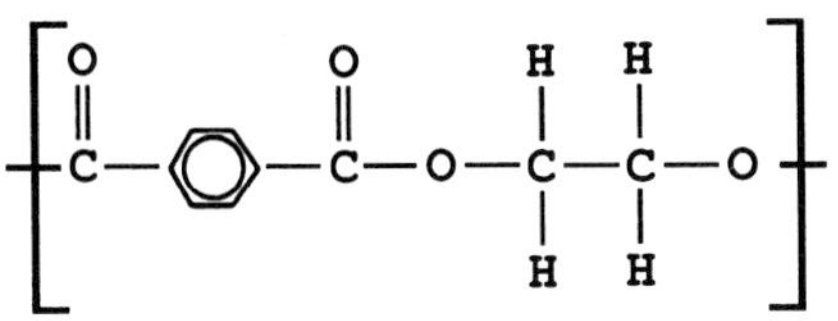

Figure 4.6 Chain repeat structure of PET.

amide is melt spun in an inert atmosphere as described in Section 4.2 , followed by drawing (with a draw ratio of 5) after cooling to obtain a high strength, thermoplastic nylon fiber. We should mention here that the use of very high speed fiber spinning machines (up to 6000 m/min) can obviate the need for the drawing step (Nakajima, 1994).

4.4.2   Polyester fibers

Polyester fibers, similar to polyamide fibers, represent another important family of fiber. Polyester fiber was discovered in England in 1941 and commercialized in 1950. Two common trade names of polyester are *Dacron* in the US and *Terylene* in the UK. The term polyester fiber represents a family of fibers made of polyethylene terephthalate. Dimethyl terephthalate is reacted with ethylene glycol in the presence of a catalyst, antimony oxide, to produce polyethylene terephthalate or polyester. The chain repeat structure of PET is given in Fig. 4.6. Although polyesters can be both thermosetting and thermoplastic, the term polyester has become synonymous with PET. Note that the PET chain structure is different from the simpler structure of nylon or polyethylene. In PET, the aromatic ring and its associated C–C bonds provide a rigidity to the structure. The polyester structure is also bulkier than that of nylon or polyethylene. These factors make polyester less flexible than nylon and polyethylene, and the crystallization rate of PET slower than that of nylon or polyethylene. Thus, when polyester is cooled from the melt, an appreciable amount of crystallization does not result.

Like nylon, polyester fibers are made from linear-condensation polymers by melt spinning, followed by drawing. Similar to nylon, the drawing treatment involves a stretch ratio of 5. The drawing of polyester fiber is done above its glass transition temperature of 80°C.

4.4.3   Polyolefin fibers

Polypropylene and polyethylene are perhaps the two most important polyolefin fibers. Polyethylene has a simple, linear chain structure consisting of a carbon backbone and small hydrogen side groups. Such a structure makes it easy to crystallize. There are three common grades of polyethylene; low density polyethylene (LDPE), high density polyethylene (HDPE), and ultra-high molecular

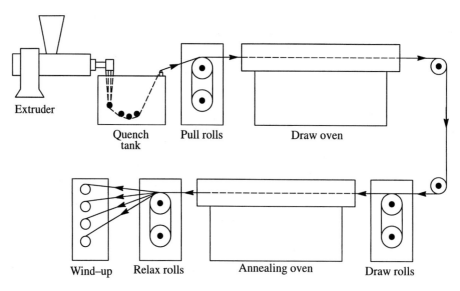

Figure 4.7   Schematic of the melt spinning process of making polypropylene fibers.

weight polyethylene (UHMWPE). It is the ultra-high molecular weight poly-ethylene variety that is used to make high modulus fibers. We describe the UHMWPE polyethylene fiber in some detail in Section 4.5 because of its com-mercial importance as a high strength and high modulus fiber. Polypropylene is an important fiber, although it does not possess a very high modulus. The degree of crystallinity achievable in polypropylene is generally less than that in poly-ethylene. This is because polypropylene has side groups while polyethylene is a highly linear polymer. In general, bulky side groups make it difficult to obtain an orderly arrangement of molecular chains, i.e. crystallization. The spatial arrangement of groups (tacticity) is also important in this regard. Polyolefin fibers are spun from polymers or copolymers of olefins such as ethylene and pro-pylene. Polypropylene fibers are made by melt spinning, involving extrusion through an extruder followed by thermal and mechanical treatment. Figure 4.7 shows a schematic of this process. The extruder has a spinneret at the one end. As in other cases, the spinneret has orifices generally arranged in a circular fashion. The melt temperature is about 250°C. The filaments go through a water-filled quench tank to pull rolls and then on to a draw oven, followed by draw rolls, an annealing oven, and finally to a windup drum. As with any other polymer, the degree of chain orientation in the polymer is a function of the draw ratio. In this case, the draw ratio is simply the ratio of the speed of the draw roll to that of the pull rolls. Commonly, this draw ratio is 9, which gives a high strength fiber. The annealing treatment serves to relieve any processing induced residual stresses.

4.4.4   Polydiacetylene single crystal fiber

We discussed the concept of degree of crystallinity in a polymer and the fact that it is very difficult to obtain a 100% crystalline polymer. Under careful conditions, it is possible to produce a fully crystalline or even a single crystal polymer. In particular, polydiacetylene single crystal fibers have been obtained by solid state polymerization (Young, 1989). The process involves preparation of single crystals of monomer by vapor phase deposition or precipitation from a dilute solution. The monomer is subjected to solid state polymerization (homogeneously or heterogeneously) by means of synchrotron radiation, gamma-rays, or thermally.

Fibers produced in this manner have some shortcomings. For example, these fibers are of very short length, no more than 5–10 cm long, and are produced rather slowly. One should also add that the only single crystal fiber produced is based on solid state polymerization of certain substituted diacetylenes and the technique does not seem to be applicable to other systems.

4.4.5   Elastomeric fibers

Elastomers can be described, in a very general way, as polymers that exhibit a rubbery behavior, i.e. they exhibit nonlinear elastic behavior and are capable of being stretched extensively. The American Society for Testing and Materials (ASTM) provides a more precise definition of an elastomer; an elastomer is 'a material which at room temperature can be stretched repeatedly to at least twice its original length, and, upon immediate release of the stretch, will return with force to its approximate original length'. There are many segmented block copolymers (see Chapter 3) that fit this requirement of being able to be stretched 200% and thus can be called elastomers. However, not all of these can be converted into a fibrous form. In short, elastomers are materials that show a very high degree of elastic, *nonlinear* strain. Most elastomers have a very low Young's modulus, E. Elastomeric materials, although they have rather low stiffness, possess very good damping properties. Their main characteristic is that they have a long range of nonlinear, reversible elastic strain. For example, an elastomer can be *elastically* stressed to a strain of 700% while steel can be elastically stressed only to about 0.1% strain. This is because in metals and ceramics, elasticity corresponds to small elastic displacements of atoms from their equilibrium position. Microstructurally elastomers are long chain polymers with a glass transition temperature, $T_g <$ room temperature (293 K). Their unstrained state corresponds to a random coil structure of the macromolecular chains while in the strained state, these macromolecules become extended chains. Thus, straining results in chain orientation which results in a decrease in entropy, S. Entropy is a measure of disorder in the chain configuration. We can write the free energy change on straining as

$$\Delta G = \Delta H - T \Delta S$$

where ΔH is the change in enthalpy, T is the temperature, and ΔS is the change in entropy. The enthalpy does not change when an elastomer is strained. Therefore, the free energy change resulting from straining an elastomeric band can be written as:

$$\Delta G = -T \Delta S$$

Consider a polymer chain extending between two points. The flexible chain can have many configurations between these points. For the configurational entropy, we can write an expression from statistical mechanics,

$$S_{config} = k \ln p$$

where k is Boltzmann's constant and p is the number of configurations. When we stretch a rubber band, we move the two fixed points further apart and the number of chain configurations possible will be reduced compared to the number of configurations possible in the unstrained state. This means that the entropy, S, is reduced on straining. Since T is always positive and ΔS is negative on straining, we have a positive ΔG. That is to say that the strained state of a rubber band is thermodynamically not favored! That is why when we release a strained rubber band (a high entropy state), the unstrained state is regained.

The essential characteristic of an elastomeric fiber is that the material can be stretched several times its original length and when unloaded, it returns to its original dimensions. This characteristic can be imparted to a polymer in two ways:

(i) Chemically cross-link flexible polymeric chains. For example, this is done by vulcanization of rubber by sulfur.
(ii) Make a two-phase, multiblock polymer, more commonly known as a segmented polymer. An example of such a polymer is the segmented polyurethanes.

Elastomeric fibers can be made out of natural rubber. Rubber latex, which is a suspension of rubber spheres in water, is mixed with a vulcanizing agent, and the mixture is wet spun. The fibers from the spinneret are extruded into a coagulating bath and then vulcanized to produce the cross-linking of chains. Such fibers are used in woven and knitted fabrics, and more often than not they are mixed with other textile fibers to form fabrics for garments. Most natural rubber fibers, however, have been supplanted by synthetic elastomeric fibers, especially in the area of athletic wear. This is attributable to the poor dyeability, abrasion resistance, and chemical stability of rubber fiber *vis-à-vis* the segmented polyurethane fibers. The term *Spandex* (or *Lycra*) is used as a generic term to denote synthetic fibers that consist of at least 85% of segmented polyurethane polymer. The term *elastane* is more common in Europe.

Processing of elastomeric fibers

Elastomeric fibers such as Spandex (or Lycra) are polyurethane fibers and are obtained by spinning from polymers in which molecular linkage occurs through urethane groups. The processing involves a series of chemical reactions, starting with the production of a low MW polymer (prepolymer), followed by reaction of the prepolymer with diisocyanate. The isocyanate-terminated prepolymer is then coupled to form the segmented polyurethane. Soluble polyurethanes that are essentially linear can be dissolved in any number of solvents and subjected to wet or dry spinning. As described above, in the wet spinning process, the fibers from the spinneret pass through a coagulating bath while in the dry spinning process they pass through an atmosphere that removes the solvent. Dry spinning is by far the most common method of making elastomeric fibers. The spinning dope has a variety of additives such as TiO_2 for delustering, stabilizers, antioxidants, lubricants, antitacking agents, dye stuff receptive agents, etc. The dope is extruded through a spinneret and the solvent is removed by flowing hot gases. Single filaments are coalesced to form a fused multifilament assembly. The spinning rates are from 200–800 m/min. Generally, a post-spinning stretching treatment is given at 150°C to obtain enhanced orientation of the hard segments parallel to the fiber axis and thus enhanced strength.

If the polyurethane molecule is allowed to form a three-dimensional structure, it becomes insoluble and neither wet nor dry spinning processes can be used. A chemical spinning process must be used in this case. The trick is to spin the isocyanate-terminated prepolymer at a stage where it forms a viscous melt, and with the jets produced in an environment that contains a chain extender. This chain extender diffuses into the fibers and reacts to couple the prepolymer molecules (Couper, 1985).

Structure and properties of elastomeric fibers

Polyurethanes that form the basis of these elastomeric fibers are made by reacting glycols and diisocyanates. Such elastomeric fibers are block copolymers. Their main characteristic is, as we said above, that they can be stretched repeatedly. Structurally, these block copolymers consist of a soft and flexible chain segment (random-coiled aliphatic polyethers or copolyesters) and a hard segment (aromatic-aliphatic polyureas). Urethane linkages connect the hard segment blocks to the soft segments. Figure 4.8a shows schematically how an elastomeric fiber consisting of hard and soft segments can show high stretchability. The soft

Soft segments   Hard segments

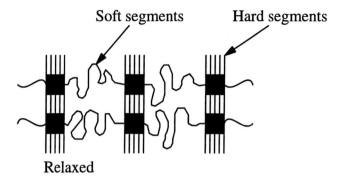

Relaxed

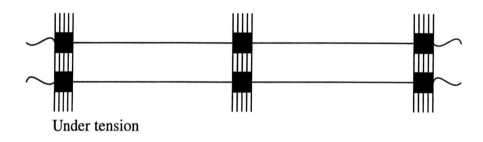

Under tension

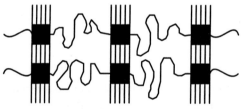

Recovery after removal of tension

Figure 4.8   (a) An elastomeric fiber consists of hard and soft chain segments and can show high stretchability.

segment domains, comprising 65–90% by weight, are unoriented in the relaxed state of the fiber but straighten out when stretched. The soft segments allow the polyurethane to stretch while the hard segments act as anchors that allow the fibers to recover to their original shape when the load is removed. One can control the amount of stretching by controlling the ratio of hard to soft segments. Spandex fibers have linear density between 1 and 500 tex and a strain to failure in the range 400–800%. Figure 4.8b shows specific stress–strain curves of Lycra (Spandex) and natural rubber fibers (Wilson, 1967, 1968). Spandex or Lycra fiber

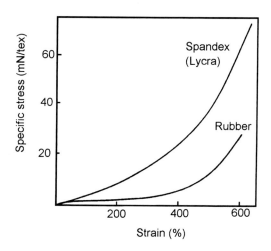

Figure 4.8 (*cont.*) (b) Specific stress–strain curves of Lycra (Spandex) and natural rubber fibers (Wilson, 1967, 1968).

is stronger than rubber fibers. It has a lower modulus than nylon but it can be stretched to much higher values of strain.

Spandex-type polyurethane fibers are chemically quite stable under normal washing and chlorine containing swimming pool water conditions. Special stabilizers are required for protection against ultraviolet radiation, suntan lotions, human perspiration, etc.

4.5   Strong and stiff polymeric fibers

As described in Chapter 3 and above, in general, macromolecular polymers have the chains in a random coil configuration, i.e. they have the so-called *cooked-spaghetti* structure. In this random coil structure, the macromolecular chains are neither aligned in one direction nor stretched out. Thus, they have predominantly weak van der Waals interactions rather than strong covalent interactions, resulting in a low strength and stiffness. Since the covalent carbon–carbon bond is a very strong bond, one would expect that linear chain polymers such as polyethylene would be *potentially* very strong and stiff. Conventional, isotropic polymers show a Young's modulus, E, of about 10 GPa. Highly drawn polymers having a Young's modulus of about 70 GPa are available commercially. The most important message of the extensive work done in the second half of the twentieth century in the area of structure–property relations in polymers is that if one wants strong and stiff organic fibers (e.g. polyethylene), then one must obtain *oriented molecular chains with full extension*. Thus, in order to obtain high stiffness and strength polymers, we must extend these polymer chains and pack them in a parallel array. The orientation of these polymer chains with respect to the fiber axis and the manner in which they fit together (i.e. order or crystallinity) are controlled by their chemical nature and the processing route.

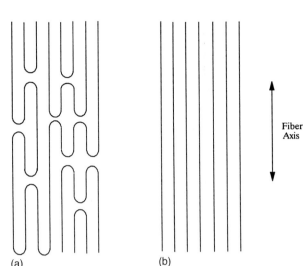

Figure 4.9 (a) Molecular orientation without high molecular extension. (b) Molecular orientation with high molecular extension.

Fiber Axis

(a)

(b)

There are two ways of achieving molecular orientation, one without high molecular extension, Fig. 4.9a, and the other with high molecular extension, Fig. 4.9b. It is the kind of chain structure shown in Fig. 4.9b, i.e. molecular chain orientation coupled with molecular chain extension, that is needed for high stiffness and strength. In the polymer fiber literature sometimes the term *ultra-high modulus* fibers is used for fibers with extremely high modulus that results from an oriented and extended chain structure. What is really implied here is a Young's modulus greater than 70 GPa, which is the modulus of aluminum and glass. To obtain a Young's modulus value greater than 70 GPa requires high draw ratios, i.e. a very high degree of elongation must be carried out under such conditions that macroscopic elongation results in a corresponding elongation at a molecular level. It turns out that the Young's modulus, E, of a polymeric fiber increases linearly with the deformation ratio (draw ratio in tensile drawing or die drawing and extrusion ratio in hydrostatic extrusion). The drawing behavior of a polymer is sensitive to:

(i) molecular weight and molecular weight distribution;

(ii) deformation conditions (temperature and strain rate).

Too low a drawing temperature produces voids, while too high a drawing temperature results in flow drawing, i.e. the macroscopic elongation of the material does not result in a molecular alignment, and consequently, no stiffness enhancement results. An oriented and extended macromolecular chain structure, however, is not easy to achieve in practice in flexible polymers. Nevertheless, considerable progress in this area has been made during the last quarter of the twentieth century. Organic fibers such as aramid and polyethylene possessing high strength and modulus are the fruits of this work. Two very different approaches have been taken to make high modulus organic fibers. These are:

(i) Processing of the conventional flexible-chain polymers in such a way that the internal structure takes a highly oriented and extended-chain arrangement. Structural modification of 'conventional' polymers such as high modulus polyethylene was developed by choosing appropriate molecular weight distributions, followed by drawing at suitable temperatures to convert the original folded-chain structure into an oriented, extended chain structure. It should be pointed out that Mark (1936) predicted way back in 1936 that the theoretical modulus of polyethylene with aligned chains should approach 250 GPa.

(ii) The second approach, radically different, involves synthesis, followed by extrusion of a new class of polymers, called liquid crystal polymers. These have a rigid-rod molecular chain structure. The liquid crystalline state, as we shall presently see, has played a very significant role in providing highly ordered, extended chain fibers.

Let us consider these two approaches in some detail. Recall that stretching polymers at temperatures below their melting point improves their stiffness and strength. Techniques involving some stretching are used commercially for nylon, polyester, and polypropylene fibers (as well as some polymer films) (see Section 4.4). Fairly high strength can be obtained but modulus is low compared to glass or aluminum. The two approaches mentioned above have resulted in two commercialized high strength and high stiffness fibers; polyethylene and aramid. Next we describe the processing, structure, and properties of these two fibers.

4.5.1 Oriented polyethylene fibers

The ultra-high molecular weight polyethylene fiber is a highly crystalline fiber with very high stiffness and strength. All of this results from some innovative processing and control of structure of polyethylene.

Processing

Drawing of melt crystallized polyethylene (molecular mass 10^4 to 10^5) to very high draw ratios can result in moduli of up to 70 GPa. Tensile drawing, die drawing or hydrostatic extrusion can be used to obtain the high permanent or plastic strains required to obtain a high modulus. It turns out that modulus is dependent on the draw ratio, but independent of how the draw ratio is obtained (Capaccio *et al.*, 1979). In all these drawing processes, the polymer chains become merely oriented without undergoing molecular extension, and we obtain the kind of structure shown in Fig. 4.9a.

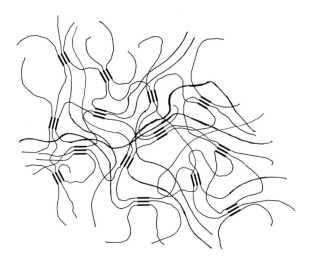

Figure 4.10 Schematic of a gel structure with crystalline regions at the network junctions.

Later developments involving solution and gel spinning of very high molecular mass polyethylene ($>10^6$) resulted in moduli as high as 200 GPa. In these methods, molecular orientation is achieved together with chain extension, i.e. the molecular chain structure shown in Fig. 4.9b is obtained. Lamellar single crystals of polyethylene can be obtained from an appropriate solution by crystallization, aligning the molecular chains in the direction of flow. In another method, a crystallized shish kebab structure is obtained from the polymer solution. A *shish kebab* structure consists of a continuous array of fibrous crystals, the shish kebabs, in which the molecular chains are highly extended.

The method that has become technologically and commercially most successful involves gel spinning of polyethylene fibers. Pennings and colleagues (1972, 1976) made high modulus polyethylene fiber by solution spinning. Their work was followed by Smith and Lemstra (1976, 1980) who made polyethylene fiber by gel spinning. The gel spinning process of making polyethylene was industrialized in the 1980s. Gels are nothing but swollen networks in which crystalline regions form the network junctions (Fig. 4.10). Essentially, an appropriate polymer solution is converted into gel which can be processed by a variety of methods to give the fiber. The aligned and extended chain structure is obtained by drawing the gelled fiber. At least three commercial firms produce oriented polyethylene fiber using very similar techniques. DSM (Dutch State Mines), produces a fiber called *Dyneema*; AlliedSignal, a US company produces *Spectra* fiber under license from DSM; while Mitsui, a Japanese company, produces a polyethylene fiber with the trade name *Tekmilon*.

Gel-spinning of polyethylene fiber

Polyethylene (PE) is a particularly simple, linear macromolecule, with the following chemical formula

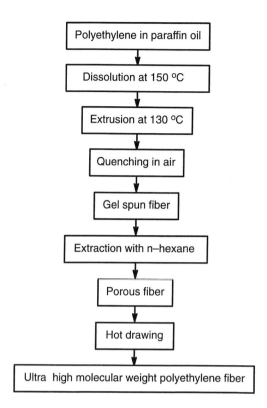

Figure 4.11 Flow diagram of the gel-spinning process of making ultra high molecular weight polyethylene fiber.

$$-[-CH_2-CH_2-]-$$

Thus, compared to other polymers, it is easier to obtain an extended and oriented chain structure in polyethylene. High density polyethylene (HDPE) is preferred to other types of polyethylene because HDPE has fewer branch points along its backbone and a high degree of crystallinity. These characteristics of linearity and crystallinity are important to obtain a high degree of orientational order and an extended chain structure in the final fiber.

Figure 4.11 provides a flow diagram of the gel-spinning process of making high modulus polyethylene fiber. The three companies mentioned earlier use different solvents (decalin, paraffin oil and paraffin wax) to make a dilute (2–10 %) solution of polymer in solvent at about 150°C. A dilute solution ensures less chain entanglement, which makes it easier for the final fiber to be highly oriented. A polyethylene gel is produced when the solution coming out of the spinneret is quenched by air. The as-spun gelled fiber enters a cooling bath. At this stage the fiber is thought to have a structure consisting of folded chain lamellae with solvent between them and a swollen network of entanglements. These entanglements allow the as-spun fiber to be drawn to very high draw ratios; draw ratios can be as high as 200. The maximum draw ratio is related to the average

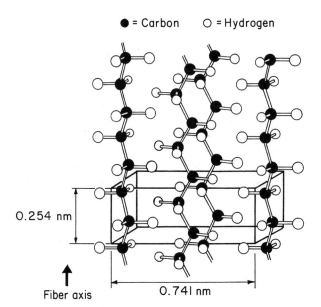

Figure 4.12 Unit cell of a single crystal of polyethylene (after Bunn, 1939; Swan, 1962). The sides of the unit cell have dimensions 0.741, 0.494 and 0.254 nm.

distance between the entanglements, i.e. the solution concentration. The gelled fibers are drawn at 120°C. One problem with this gel route is the rather low spinning rates of 1.5 m min^{-1}. At higher rates, the properties obtained are not very good (Kalb and Pennings, 1980; Smook and Pennings, 1984).

Structure and properties of polyethylene fiber

The crystal structure of polyethylene has been studied extensively. Figure 4.12 shows the unit cell of a single crystal of polyethylene (Bunn, 1939; Swan, 1962). The sides of the unit cell have the dimensions 0.741, 0.494 and 0.255 nm. There are four carbon and eight hydrogen atoms per unit cell. One can then compute the theoretical density of polyethylene (i.e. assuming a 100% single crystal polyethylene) to be 0.9979 g cm^{-3}. Of course, in practice, one can only tend toward this theoretical value. As it turns out, the highly crystalline UHMWPE fiber has a density of 0.97 g cm^{-3}; very close to the theoretical value. Thus, polyethylene fiber is very light; in fact, it is lighter than water and thus floats on water. A summary of some commercially available polyethylene fibers is provided in Table 4.1.

Its strength and modulus are slightly lower than those of aramid fibers on a per-unit-weight basis, i.e. specific property values are about 30–40% higher than those of aramid. It should be pointed out that both polyethylene and aramid fibers, as is true of most organic fibers, must be limited to low temperature (less than 150°C) applications.

Another effect of the high degree of chain alignment in these fibers is manifested when they are put in a polymeric matrix to form a fiber reinforced composite. High modulus polyethylene fibers such as Spectra or Dyneema are hard to bond with any polymeric matrix. Some kind of surface treatment must be given to the polyethylene fiber for it to bond with resins such as epoxy, PMMA, etc. By far the most successful surface treatment involves a cold gas (air, ammonia, argon, etc.) plasma (Kaplan *et al.*, 1988). A plasma consists of gas molecules in an excited state, i.e. highly reactive, dissociated molecules. When polyethylene, or any other fiber, is treated with a plasma, surface modification occurs by removal of any surface contaminants and highly oriented surface layers, addition of polar and functional groups on the surface and introduction of surface roughness, all these factors contribute to enhanced fiber/matrix interfacial strength (Biro *et al.*, 1992; Brown *et al.*; 1992, Hild and Schwartz, 1992a, 1992b; Kaplan *et al.*, 1988; Li *et al.*, 1992). An exposure of just a few minutes to the plasma is sufficient to do the job!

Commercially available polyethylene fiber has a degree of crystallinity between 70 and 80% and a density $0.97\,\mathrm{g\,cm^{-3}}$. There is a linear relationship between density and crystallinity for polyethylene. A 100% crystalline poly ethylene will have a theoretical density, based upon an orthorhombic unit cell, of about $1\,\mathrm{g\,cm^{-3}}$. A totally amorphous polyethylene (0% crystallinity) will have a density of about $0.85\,\mathrm{g\,cm^{-3}}$. Khosravi *et al.* (1995) used nitric acid attack on gel-spun polyethylene fibers to observe structural imperfections such as fold, molecular kinks and uncrystallized regions. Raman spectroscopy has been used to study the deformation behavior of polyethylene fiber. This technique gives peaks for the crystalline and amorphous states of polyethylene (see Chapter 9).

4.5.2   Aramid fibers

The second approach mentioned above to making high stiffness and high strength polymeric fiber is the *liquid crystal route*. This involves synthesis and

Table 4.1   *Properties of polyethylene fibers.*[a]

Property	Spectra 900	Spectra 1000
Density (g cm^{-3})	0.97	0.97
Diameter (μm)	38	27
Tensile strength (GPa)	2.7	3.0
Tensile modulus (GPa)	119	175
Tensile strain-to-fracture (%)	3.5	2.7

Note:
[a] AlliedSignal's data. Indicative values.

extrusion of rigid-rod molecular chain polymers. From a historical perspective, it can be said that research into the production of stiff chain aromatic poly-amides started some time in the mid-twentieth century. It was soon discovered, however, that there was a severe processing bottleneck, namely the extreme insolubility of aromatic polyamides. In 1965, Stephanie Kwolek, a research scientist at Du Pont, discovered that para aminobenzoic acid could be polymerized and solubilized, under certain conditions, to form a rigid rod liquid crystalline solution that was spinnable. That discovery can be marked as the beginning of aramid fibers. Later on it was found that the polymer obtained by reacting p-phenylene diamine and terephthalic acid was better. It should be recognized that important contributions were made by researchers at Monsanto, although Monsanto decided against commercializing an aramid fiber. We provide below a summary of the processing, structure and properties of aramid fibers.

Processing of aramid fibers

The term aramid in aramid fiber is a short form for aromatic polyamide. As described above, conventional polyamides, e.g. nylon, contain mostly aliphatic and cycloaliphatic units in the macromolecular chain structure. Aramid is a generic term that represents an important class of fibers. The United States Federal Trade Commission defines aramid as 'a manufactured fiber in which the fiber forming substance is a long chain synthetic polyamide in which 85% of the amide linkages are attached directly to two aromatic rings'. Commercial names of aramid fibers include Kevlar and Nomex (Du Pont); Teijinconex and Technora (Teijin); and Twaron (Akzo). The basic difference between Kevlar and Nomex is that Kevlar has *para*-oriented aromatic rings, i.e. it has a symmetrical molecule, with bonds from each aromatic ring being parallel, while Nomex is *meta*-oriented, with bonds at an angle of 120° to each other.

Perhaps it is worth repeating and clarifying some of the nomenclature information at this point. Aramid fiber is a generic name of a class of synthetic organic fibers called aromatic polyamide fibers. Nylon is a generic name for any long chain polyamide containing less than 85% aromatic groups. Aramid fibers such as Nomex or Kevlar are ring compounds based on the structure of benzene, as opposed to linear compounds used to make nylon. The basic chemical structure of Kevlar aramid fibers consists of oriented para-substituted aromatic units which limits conformational freedom and makes them rigid rodlike polymers. This rigid rodlike structure results in a high glass transition temperature and poor solubility, which makes fabrication of these polymers, by conventional drawing techniques difficult. Instead, they are spun from liquid crystalline polymer solutions as described below. Many aromatic polymers can be converted to high modulus fibers via the liquid-crystal route. In theory, it was expected that such a rigid-rod molecular structure should result in a highly oriented fiber structure in the as-spun state, i.e. without any need to resort to drawing after spinning. There were, however, two practical problems:

(i) These aromatic polyamides have relatively high melting points or they degrade thermally at high temperatures. Thus, they could not be melt spun.

(ii) The viscosity of their isotropic solutions is very high which renders them unsuitable for spinning.

It was, however, observed that such systems under appropriate conditions of concentration, solvent, molecular weight, temperature, etc. form a liquid crystalline solution. Perhaps a little digression is in order here to say a few words about liquid crystals. A liquid crystal has a structure intermediate between a three-dimensionally ordered crystal and a disordered isotropic liquid. There are two main classes of liquid crystals: *lyotropic* and *thermotropic.* Lyotropic liquid crystals are obtained from low viscosity polymer solutions in a critical concentration range while thermotropic liquid crystals are obtained from polymer melts where a low viscosity phase forms over a certain temperature range. Aromatic polyamides and aramid type fibers are lyotropic liquid crystal polymers. These polymers have a melting point that is high and close to their decomposition temperature. One must therefore spin these from a solution in an appropriate solvent such as sulfuric acid. Aromatic polyesters, on the other hand, are thermotropic liquid crystal polymers. These can be injection molded, extruded or melt spun.

There are many types of liquid crystals. Figure 4.13 shows the structure of some of these. All of them contain ordered domains. The main difference is in the orientation of these ordered domains in the liquid. It is the nematic variety that is of interest to us here. In order to see under what conditions one can obtain a liquid crystal, a phase diagram indicating different forms present under different conditions is of great help. For example, Fig. 4.14a shows a typical phase diagram for a poly (p-phenylene terephthalamide)–sulfuric acid system, a solution that is used to make Kevlar fiber. The key to processability of aramid fibers is that the viscosity of an isotropic solution increases with polymer concentration *until* an anisotropic phase starts separating out, at which point the viscosity shows a marked drop as shown in Fig. 4.14b. Solutions of rodlike polymers show a spontaneous transition between isotropic and nematic liquid crystal phase above a critical polymer phase concentration. Actually, with hindsight we can say that Flory's (1956) statistical theory predicts this transition, i.e. the onset of solution anisotropy in liquid crystal solutions at a critical concentration of polymer. His theory tackles the problem of packing the rigid rodlike molecules in a given volume.

Recall that conventional polyamides (e.g. nylon) contain mostly aliphatic and cycloaliphatic units in the macromolecular chain structure. These can be melt

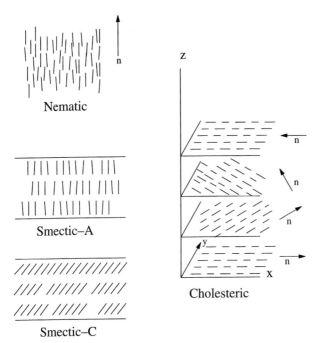

Figure 4.13 Different types of liquid crystals.

spun into fibers. Processing of aramid fibers involves solution polycondensation of diamines and diacid halides at low temperatures (Hodd and Turley, 1978; Morgan, 1979; Magat, 1980; Panar *et al.*, 1983), for example, low temperature polycondensation of *p*-phenylene diamine (PPD) and terephthaloyl chloride (TCl) in a dialkyl amide solvent. The amide solvents used are N-methyl pyrrolidone and dimethyl acetamide, separately or mixed, and generally in the presence of inorganic salts such as LiCl or $CaCl_3$. The polymer is precipitated with water, neutralized, washed, and dried. Poly p-phenylene terephthalamide (PPTA) polymer is insoluble in ordinary solvents but dissolves in strong acids such as concentrated sulfuric acid. The most important prerequisite is that the starting spinnable solution have a nematic liquid crystalline order, which results in high strength and high modulus fibers. Figure 4.15 schematically shows various states of a polymer in solution. Figure 4.15a shows two-dimensional, linear, flexible chain polymers in solution. These are called random coils as the figure suggests. If the polymer chains can be made of rigid units, i.e. rodlike, we can represent them as a random array of rods, Fig.4.15b. Any associated solvent may contribute to the rigidity and to the volume occupied by each polymer molecule. With increasing concentration of rodlike molecules, one can dissolve more polymer by forming regions of partial order, i.e. regions in which the chains form a parallel array. This partially ordered state is called a liquid crystalline state, Fig. 4.15c. When the rodlike chains become approximately arranged in parallel to their long axes but their centers remain unorganized or randomly distributed, we have what

(a)

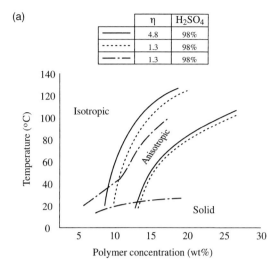

η	H₂SO₄
4.8	98%
1.3	98%
1.3	98%

Figure 4.14   (a) A typical phase diagram for a poly (p-phenylene terephthalamide)–sulfuric acid system, a solution that is used to make Kevlar fiber (after Kikuchi, 1982). (b) Viscosity versus polymer concentration in PPTA/H$_2$SO$_4$ solution. Note the sharp drop in viscosity at about 20% which corresponds to a transition between isotropic and nematic liquid crystal phase.

(b)

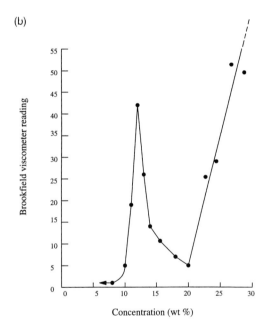

is called a nematic (Greek for *threaded*) liquid crystal, Fig. 4.15d. It is this kind of order which is found in the extended chain polyamides.

Liquid crystals, due to the presence of the ordered domains, are optically anisotropic, i.e. birefringent. This can be easily verified by observing the liquid crystal, e.g. aramid and sulfuric acid solution, at rest between crossed polarizers. The parallel arrays of polymer chains in liquid crystalline state become even more ordered when these solutions are subjected to shear as, for example, in extruding through a spinneret hole. It is this inherent property of liquid crystal

(a)

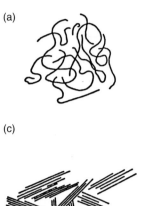

(b)

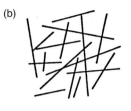

Figure 4.15 (a) Two-dimensional, linear, flexible chain polymers in solution. (b) Random array of rods. (c) Liquid crystalline state. (d) Nematic liquid crystal.

(c)

(d)

solutions which is exploited in the manufacture of aramid fibers. The characteristic fibrillar structure of aramid fibers is due to alignment of polymer crystallites along the fiber axes. Para-oriented aromatic polyamides form liquid crystal solutions under certain conditions of concentration, temperature, solvent and molecular weight. Such a liquid crystal shows the anomalous relationship between viscosity and polymer concentration described above. Initially, there occurs an increase in viscosity as the concentration of polymer in solution increases, as would be expected in any ordinary polymer solution. At a critical point where the solution starts assuming an anisotropic liquid crystalline shape, there occurs a sharp drop in the viscosity. The liquid crystalline regions act like dispersed particles and do not contribute to solution. With increasing polymer concentration, the amount of liquid crystalline phase increases up to a point when the viscosity tends to rise again. There are other requirements for forming a liquid crystalline solution from aromatic polyamides. Molecular mass must be above some minimum value and solubility must exceed the critical concentration required for liquid crystallinity. Thus, starting from liquid crystalline spinning solutions containing highly ordered arrays of extended polymer chains, fibers can be spun directly into an extremely oriented, chain extended form. These as-spun fibers are quite strong, and as the chains are highly extended and oriented, one does not need to use conventional drawing techniques. Any such post-drawing stretching will, of course, result in some improvement in the chain alignment. Para-oriented rigid diamines and dibasic acids give polyamides which yield, under appropriate conditions of solvent, concentration and polymer molecular weight, the desired nematic liquid crystal structure. One would like to have, for any solution spinning process, a high molecular weight in order to have improved mechanical properties, a low viscosity to ease processing conditions, and a high polymer concentration for high yield. For para-aramid, poly *p*-pheny-

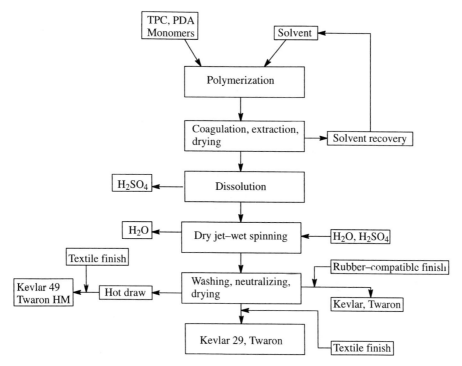

Figure 4.16 Flow diagram for making different types of aramid fibers (Ohta, 1983).

lene terephthalamide (PPTA), trade name Kevlar, the nematic liquid crystalline state is obtained in 100% sulfuric acid at a polymer concentration of about 20%. The polymer solution, often referred to as the dope, has concentrated sulfuric acid as a solvent for PPTA. Five moles of sulfuric acid are needed per PPTA amide bond, which translates into about 4 kg of sulfuric acid per kg of polymer. The spent acid after spinning is converted to calcium sulfate (gypsum). For every kg of fiber, 7 kg of gypsum is produced. Figure 4.16 shows the flow diagram for making different types of aramid fibers (Kevlar, Twaron, etc.) (Ohta, 1983).

For aramid fibers, the dry jet–wet spinning method is employed. The process is illustrated in Fig. 4.17. Solution-polycondensation of diamines and diacid halides at low temperatures (near 0°C) gives the aramid-forming polyamides. Low temperatures are used to inhibit any by-product generation and promote linear polyamide formation. The resulting polymer is pulverized, washed and dried, mixed with concentrated H_2SO_4 and extruded through a spinneret at about 100°C. The jets from the orifices pass through about 1 cm of air layer before entering a cold water (0–4°C) bath. The fiber precipitates in the air gap, and the acid is removed in the coagulation bath. The spinneret capillary and air gap cause alignment of the domains resulting in highly ordered, crystalline and oriented as-spun fibers. The air gap also allows the dope to be at a higher temperature than

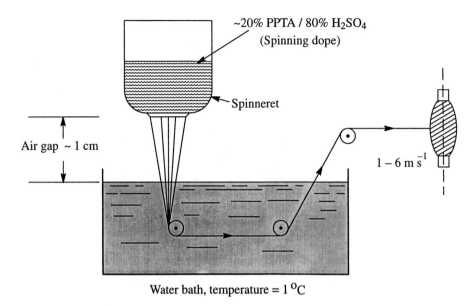

~20% PPTA / 80% H$_2$SO$_4$
(Spinning dope)

Spinneret

Air gap ~ 1 cm

1 – 6 m s^{-1}

Water bath, temperature = 1 $^{\circ}$C

Figure 4.17   Schematic of the dry jet–wet spinning method.

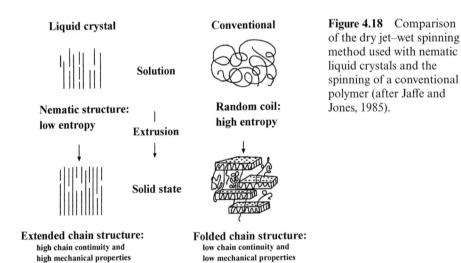

Liquid crystal

Solution

Conventional

Nematic structure:
low entropy

Extrusion

Random coil:
high entropy

Solid state

Extended chain structure:
high chain continuity and
high mechanical properties

Folded chain structure:
low chain continuity and
low mechanical properties

Figure 4.18   Comparison of the dry jet–wet spinning method used with nematic liquid crystals and the spinning of a conventional polymer (after Jaffe and Jones, 1985).

would be possible without the air gap. The higher temperature allows a more concentrated spinning solution to be used and higher spinning rates are possible. Spinning rates of several hundred meters per minute are not unusual. Figure 4.18 compares the dry jet–wet spinning method used with nematic liquid crystals and the spinning of a conventional polymer. The oriented chain structure together with molecular extension is achieved with dry jet–wet spinning. The conventional wet or dry spinning gives precursors that need further processing for a

Figure 4.19   Different forms in which Kevlar aramid fiber is available: pulp, floc, staple, and continuous fiber woven into a variety of fabrics (courtesy of Du Pont).

marked improvement in properties (Jaffe and Jones, 1985). The as-spun fibers are washed in water, wound on a bobbin and dried. Fiber properties are modified by the use of appropriate solvent additives, by changing the spinning conditions, and by means of some post-spinning heat treatments.

In addition to the continuous filament yarn, Kevlar aramid fiber is also available in several short forms such as staple, floc and highly fibrillated, pulp form. Kevlar pulp can be used in adhesion, sealants and coatings. Such products can be used to provide control in addition to the general reinforcement of resins such as epoxy. Figure 4.19 shows some of these forms of Kevlar aramid fiber. Kevlar aramid staple fiber is generally preion-cut to lengths of 6 mm or greater. It can be used to manufacture spun yarns, felts and nonwovens for thermal insulation and vibration damping purposes. Kevlar floc refers to fiber shorter than staple (~1 mm length) and can be used as reinforcement for resins.

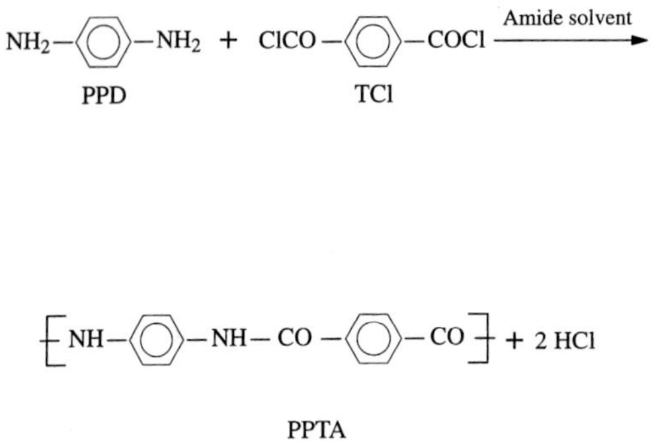

$$NH_2-⟨○⟩-NH_2 \; + \; ClCO-⟨○⟩-COCl \xrightarrow{\text{Amide solvent}}$$

PPD   TCl

Figure 4.20   Chemical synthesis formula for aramid fiber.

$$\left[NH-⟨○⟩-NH-CO-⟨○⟩-CO \right] + 2\,HCl$$

PPTA

Teijin aramid fiber, known as Technora (formerly as HM-50), is made slightly differently from the liquid crystal route described above. Three monomers, terephthalic acid, p-phenylenediamine (PDA), and 3,4-diamino diphenyl ether are used. The ether monomer provides more flexibility to the backbone chain which results in a fiber that has slightly better compressive properties than PPTA aramid fiber made via the liquid crystal route. An amide solvent with a small amount of salt (calcium chloride or lithium chloride) is used as a solvent (Ozawa *et al.*, 1978). The polymerization is done at 0–80°C in 1–5 h and with a polymer concentration of 6–12%. The reaction mixture is spun from a spinneret into a coagulating bath containing 35–50% $CaCl_2$. Draw ratios between 6 and 10 are used.

Structure

Chemically, the Kevlar or Twaron type aramid fiber is poly(p-phenylene terephthalamide) which is a poly-condensation product of terephthaloyol chloride (TCl) and p-phenylene diamine (PPD). Its chemical formula is given in Fig. 4.20. The aromatic rings impart the rigid rodlike characteristics of Kevlar. These chains are highly oriented and extended along the fiber axis with the resultant high axial modulus. Kevlar has a highly crystalline structure and the linearity of the polymer chains results in a high packing efficiency. The unit cell of Kevlar fiber is orthorhomic with the dimensions: $a=0.785\,nm$, $b=0.515\,nm$ and $c=1.28\,nm$ (Northolt, 1974). An important feature of aramid fiber is the directional bonding: strong covalent bonding in the fiber direction and weak hydrogen bonding in the transverse direction, see Fig. 4.21a. This results in highly anisotropic properties of aramid fiber. Dobbs *et al.* (1980) examined the structure of Kevlar fiber by electron microscopy and diffraction. Based upon this work, a schematic representation of the supramolecular structure of Kevlar 49 is shown in Fig. 4.21b. It shows radially arranged, axially pleated crystalline supramolecular sheets. The molecules form a planar array with interchain hydrogen bonding. The stacking sheets form a crystalline array, but between the sheets the bonding is weak. Each pleat is about 500 nm long and the pleats are separ-

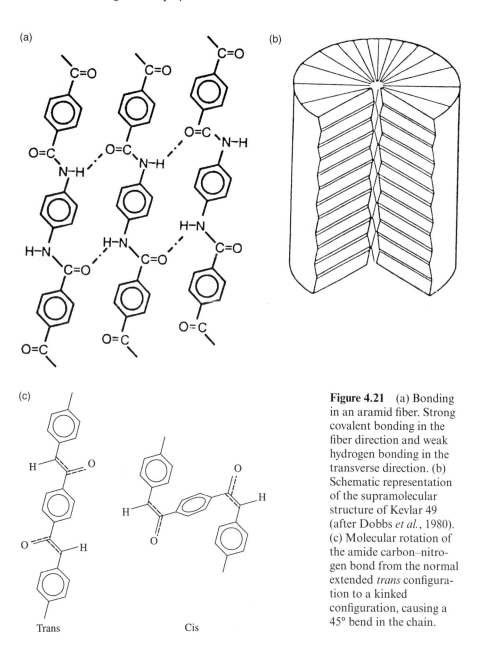

Figure 4.21 (a) Bonding in an aramid fiber. Strong covalent bonding in the fiber direction and weak hydrogen bonding in the transverse direction. (b) Schematic representation of the supramolecular structure of Kevlar 49 (after Dobbs *et al.*, 1980). (c) Molecular rotation of the amide carbon–nitrogen bond from the normal extended *trans* configuration to a kinked configuration, causing a 45° bend in the chain.

ated by transitional bands. The adjacent components of a pleat make an angle of 170°. Such a structure would appear to be consistent with the experimentally observed low longitudinal shear modulus and poor properties in compression in axial as well as transverse directions. Poor shear and compressive properties are, however, also observed in polyethylene and PBO fibers which do not show a pleated structure. A correlation between good compressive characteristics and a high glass transition temperature (or melting point) has been suggested

(Northolt, 1981; Kozey and Kumar, 1994). Thus, with the glass transition temperature of organic fibers being lower than that of inorganic fibers, the former would be expected to show poorer properties in compression. For aramid and similar fibers, compression results in the formation of kink bands leading to eventual ductile failure. Yielding is observed at about 0.5% strain. This is thought to correspond to a molecular rotation of the amide carbon–nitrogen bond from the normal extended *trans* configuration to a kinked cis configuration. This causes a 45° bend in the chain as shown in Fig. 4.21c. This bend propagates across the unit cell, the microfibrils, and a kink band results in the fiber. This anisotropic behavior of aramid fiber can be easily demonstrated by knotting the fiber and observing it in a microscope. Buckling or kink marks will be visible on the compressive side of a knotted aramid fiber. Such markings on the surface of aramid-type fibers have been reported by many researchers. DeTeresa *et al.* (1989), for example, observed kink bands in Kevlar when subjected to uniform compression or torsion. This phenomenon of low compressive strength of high performance organic fibers is very important and is described in Section 4.5.3.

Properties of aramid fibers

Some of the important properties of Kevlar aramid fibers are summarized in Table 4.2. As can be seen from this table, the Kevlar aramid fiber is very light and has very high stiffness and strength in tension. The two well-known varieties are Kevlar 49 and Kevlar 29. Kevlar 29 has about half the modulus but double the strain to failure of Kevlar 49. It is this high strain to failure of Kevlar 29 that makes it useful for making vests that are used for protection against small arms. It should be emphasized that aramid fiber, like other high performance organic fibers, has rather poor characteristics in compression, its compressive strength being only about one-eighth of its tensile strength. This follows from the anisotropic nature of the fiber as discussed above. In tensile loading, the load is carried by the strong covalent bonds while in compressive loading, weak hydrogen bonding and van der Waals bonds come into play which lead to rather easy local yielding, buckling and kinking of the fiber. Thus, aramid-type high performance fibers are not recommended for applications involving compressive forces.

Kevlar aramid fiber has good vibration damping characteristics. Dynamic (commonly sinusoidal) perturbations are used to study the damping behavior of a material. The material is subjected to an oscillatory strain. We can characterize the damping behavior in terms of a quantity called the logarithmic decrement, Δ, which is defined as the natural logarithm of the ratio of amplitudes of successive vibrations, i.e.

$$\Delta = \ln \frac{\theta_n}{\theta_{n+1}}$$

where θ_n and θ_{n+1} are the two successive amplitudes. The logarithmic decrement is proportional to the ratio of maximum energy dissipated per cycle/maximum energy stored in the cycle. Composites of Kevlar aramid fiber/epoxy matrix show about five times the loss decrement of glass fiber/epoxy.

Like other polymers, aramid fibers are sensitive to ultraviolet (UV) light. When exposed to ultraviolet light for an extended period, they discolor from yellow to brown and lose mechanical properties. Radiation of a particular wavelength can cause degradation because of absorption by the polymer and breakage of chemical bonds. According to Du Pont data, if one compares the energy absorption by Kevlar fiber and the incident energy during a typical midsummer day, the region of overlap is in the wavelength range between 300 and 400 nm (see Fig. 4.22). This spans the near-UV and part of the visible spectrum. It is recommended that this wavelength range should be avoided for outdoor applications involving the use of unprotected aramid fibers. A small amount of such light emanates from incandescent and fluorescent lamps or sunlight filtered by window glass. Du Pont Co. recommends that Kevlar aramid yarn should not be stored within one foot (0.3 m) of fluorescent lamps or near windows.

Table 4.2 *Properties of Kevlar aramid fiber yarns.*[a]

Property	K 29	K 49	K 68	K 119	K 129	K 149
Density (g cm^{-3})	1.44	1.45	1.44	1.44	1.45	1.47
Diameter (μm)	12	12	12	12	12	12
Tensile strength (GPa)	2.8	2.8	2.8	3.0	3.4	2.4
Tensile strain-to-fracture (%)	3.5–4.0	2.8	3.0	4.4	3.3	1.5–1.9
Tensile modulus (GPa)	65	125	101	55	100	147
Moisture regain (%) @ 25°C, 65% RH	6	4.3	4.3	—	—	1.5
Coefficient of expansion (10^{-6} K^{-1})	−4.0	−4.9	—	—	—	—

Note:
[a] All data from Du Pont brochures. Indicative values only. 25 cm yarn length was used in tests (ASTM D-885). K stands for Kevlar, a trade mark of Du Pont.

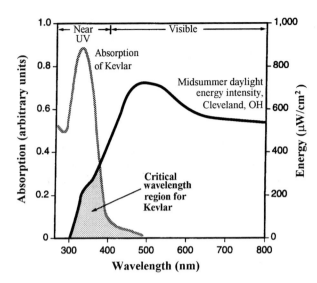

Figure 4.22 Energy absorption by Kevlar fiber and the incident energy during a typical mid-summer day (Du Pont's data). The region of overlap is in the wave-length range between 300 and 400nm.

The Technora fiber made by Teijin shows properties that are a compromise between conventional fibers and rigid-rod fibers. Table 4.3 provides a summary of these. In terms of its stress–strain behavior, it can be said that Technora fiber lies in between Kevlar 49 and Kevlar 29.

Other liquid crystal fibers

Although aramid fiber is by far the most successful fiber made via the liquid crystal route, there are some other important fibers that have been made by this process. For example, poly(p-phenylene benzobisthiazole) (PBT) (Wolfe *et al.*, 1981a, 1981b) is synthesized from terephthalic acid and 2,5-diamino-1,4-benzenedithiol dihydrochloride (DBD). The DBD is first dissolved in polyphosphoric acid (PPA), followed by dehydrochlorination. Terephthalic acid and more PPA are then added and the mixture is heated to 160°C to make a solution. The solution is heated to 180°C and reacted for 18 hours to obtain the

Table 4.3 *Properties of Technora fiber[a]*

Density (g cm^{-3})	1.39
Diameter (μm)	12
Tensile strength (GPa)	3.1
Tensile modulus (GPa)	71
Tensile strain-to-fracture (%)	4.4

Note:
[a] Manufacturer's data. Indicative values only.

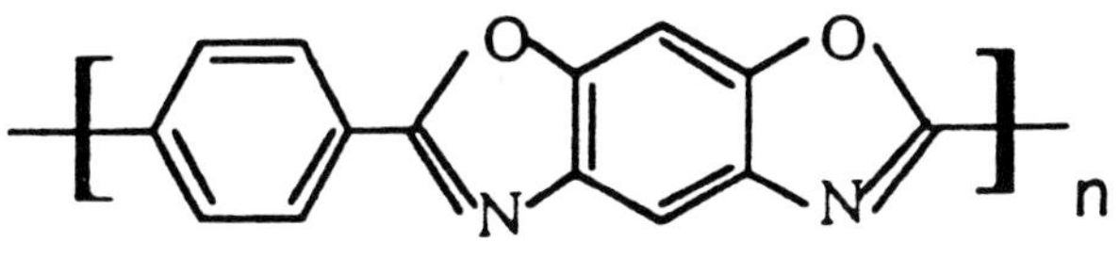

Figure 4.23 The chemical structure of PBO.

Poly(p-phenylene benzobisoxazole)

desired molecular weight. This is followed by a dry jet–wet spinning technique, similar to the one described above for spinning the aramid fibers. Drawing is done after spinning to improve the orientation to the fiber. The final step involves processing in a tubular oven in a nitrogen atmosphere. Thus, PBT is another example of a rodlike polymer made by dry jet–wet spinning of a lytropic liquid crystal. It is difficult to process but shows a very high tensile modulus.

Copolyesters combine an aromatic backbone and flexible segments. They can be melt processed. Melt processable nematic thermotropic polyesters have been produced under Vectran fiber trademark by Hoechst Celanese. Another promising rigid-rod lyotropic liquid crystal based fiber is poly(p-phenylene benzo-bisoxazole), PBO (Hokudoh et al., 1995). The chemical structure of PBO is given in Fig. 4.23. It is reported to have a tensile strength of over 5 GPa and a tensile modulus close to 300 GPa. As can be seen from these values, this fiber has better mechanical characteristics than aramid fiber. It also has higher thermal and fire resistance than aramid. Toyobo Co. in Japan has commercialized this fiber. The processing of this fiber involves a very viscous polymer solution and Toyobo Co. claims to have devised a special nozzle and control technology for fiber spinning.

4.5.3 Axial compressive strength

Highly oriented polymeric fibers such as polyethylene, aramid, etc., show superior tensile properties along the axis, but poor properties in longitudinal compression and transverse to the fiber axis. For example, the ratio of compressive strength to tensile strength can be as low as 10–20%. Such is not the case with isotropic fibers such as glass, boron and silicon carbide. PAN-based carbon fibers show better compression properties than pitch-based fibers. This discrepancy between the tensile and compressive properties has been the subject of investigation by a number of researchers (DeTeresa et al., 1982, 1984, 1985, 1988; Allen, 1987; Kozey et al., 1994; Kumar et al., 1988; Kumar and Helminiack, 1989b; Kumar, 1989, 1990a, b, 1991; Kumar and Adams, 1990; Martin and Thomas, 1989; Jiang et al., 1991a, b). Figure 4.24 shows tensile and compressive stress–strain curves of poly (p-phenylene benzobisthiazole) (PBZO) fibers under different conditions: as spun, as coagulated, and heat-treated (Martin and Thomas, 1989; Kumar, 1990a). Examples of kinking under compression in differ-

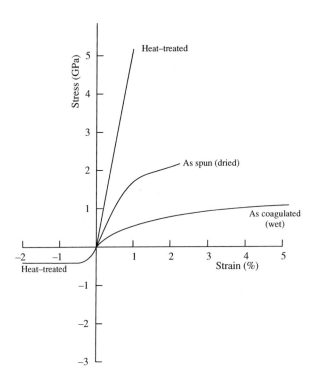

Figure 4.24   Tensile and compressive stress–strain curves of PBZO fibers under different conditions: as spun, as coagulated, and heat-treated (Martin and Thomas, 1989; Kumar, 1990a). Note the very low strength in compression.

ent high performance polymeric fibers (Kevlar 49, PBZT and PBO) are shown in Fig. 4.25. Exact mechanism(s) of failure under compression in these fibers is an open question. It would appear that the failure mechanisms in compression are different from those in tension. Various models have been proposed to explain this behavior of high performance fibers. Figure 4.26 shows two compressive failure models: (a) elastic *microbuckling* of polymeric chains; and (b) *misorientation.* The microbuckling model involves cooperative in-phase buckling of closely spaced chain in a small region of fiber. The misorientation model takes into account structural imperfections or misorientations that are invariably present in a fiber. In the composites literature it has been reported that regions of misorientation in a unidirectional composite lead to kink formation under compressive loading (Argon, 1972). The model shown in Figure 4.26b is based upon the presence of such a local misorientation in the fiber leading to kink formation under compression. Failure in compression is commonly associated with the formation and propagation of kinks. These kink bands generally start near the fiber surface and then grow to the center of the fiber. It has also been attributed to the ease of microbuckling in such fibers as well as to the presence of microvoids and the skin–core structure of these fibers. An estimate of compressive strength based upon the microbuckling model is (Schuerch, 1966; Greszczuk, 1975)

$$\sigma_{comp} = G$$

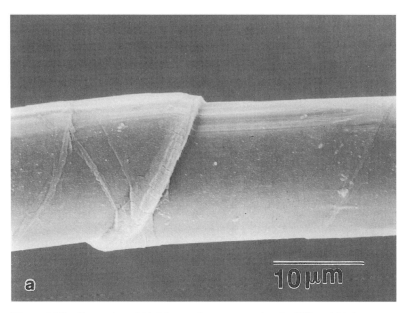

Figure 4.25 Examples of kinking under compression in different high performance polymeric fibers (a) Kevlar 49, (b,c,d) PBZT, and (e) PBO. (Courtesy of S. Kumar.) For (b, c, d) and (e) see overleaf.

where G is the longitudinal shear modulus. Experimentally, this relationship gives an overestimate. This expression assumes microbuckling failure to occur normal to the fiber axis. However, kink formation generally occurs at angles other than 90°. DeTeresa *et al.* (1984) treated a fiber under axial compression as an end loaded column on an elastic base, i.e. the axial compressive failure of these fibers occurs by elastic microbuckling instability and not by bending. They suggested an empirical relationship for the strength in compression as

$$\sigma_{comp} = 0.3\ G$$

where G is the shear modulus. Crystal misorientation away from the fiber axis occurs during compression.

Efforts to improve the compressive properties of rigid-rod polymer fibers have involved introduction of cross-linking in the transverse direction (Bhattacharya, 1989; Spillman *et al.*, 1993) and coating the fiber surface with a thin layer of a high modulus material (McGarry and Moalli, 1991, 1992).

4.6   Polymeric fibers with unusual characteristics

Polymeric fibers can have some very unusual characteristics. We describe some of these.

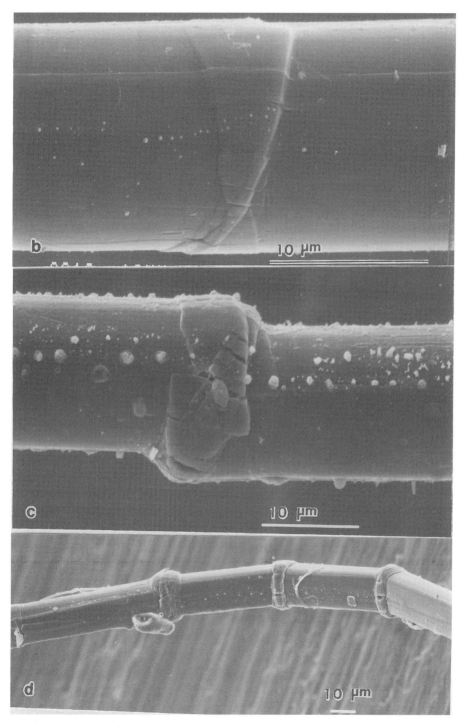

Figure 4.25   (*cont.*)

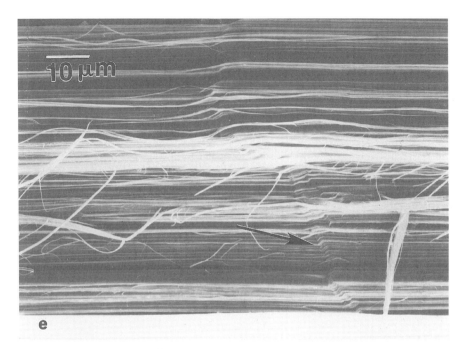

e

Figure 4.25 (*cont.*)

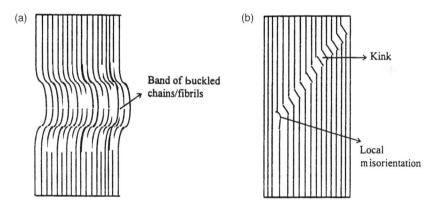

Figure 4.26 Compressive failure models in high-performance fibers: (a) micro-buckling model; and (b) misorientation model (after Kozey and Kumar, 1994).

4.6.1 Friction

In a multifilament yarn or in a braided fabric, normal (i.e. in the radial direction) frictional force holds the fibers together. Such an interfiber friction is desirable if we wish to have strong yarns and fabrics. However, there are situations where we would like to have a smooth fiber surface. For example, in a yarn passing round a guide, a smooth fiber surface is desirable. If the fiber surface is rough, then high tensile stress will occur, which, in turn, can lead to fiber break-age. In general, in textile applications, frictional characteristics can affect handle, feel, wear-resistance, etc. In fibrous composites, the frictional character-istics of fiber can affect its interface strength and toughness characteristics. Extremely smooth polymeric fibers with a very low coefficient of friction, μ (as low as 0.05) can be produced from a variety of polymers. Examples include poly(tetrafluorethylene), PTFE or Teflon as it is commonly called. PTFE is used as a coating material to produce non-stick surfaces. Highly oriented poly-meric fibers such Spectra and Dyneema are also very smooth, which makes it difficult to have them bond well to polymeric matrix materials in a composite. Generally, some kind of plasma surface treatment is used to overcome this problem.

4.6.2   Static electricity and conductive polymeric fibers

Static electricity in fibers can cause a number of problems, during processing as well as in service. The phenomenon is common in polymeric fibers. There can be two types of problem. One problem results when we have like charges. Like charges repel, which can make handling difficult. For example, filaments in a charged yarn will tend to bow out. Fabrics made with fibers that accumulate static electricity will not stay in place. Most of us have experienced this problem when wearing clothes made of synthetic fibers, especially in dry weather. The second type of problem results when we have unlike charges. Unlike charges, as we know, attract. This can cause difficulties in the opening of parachutes, garment sticking, and attraction of oppositely charged dirt and soot particles to a charged body. It should be pointed out that there is always a preponderance of *negatively* charged dust particles. Thus, soiling of garments is worse when the fabric is positively charged because it attracts negatively charged dust particles from the atmosphere. Natural fibers such as cotton rarely give static while some other natural fibers such as wool and silk do show some static problems. Synthetic fibers such as nylon, PET, aramid, etc. can show extensive charge accumulation. The problem of static can be attenuated by increasing humidity by simply moistening the fabric.

Normally one thinks of polymers and polymeric fibers as electrical insulators. However, there is some laboratory-scale work that has resulted in electrically conducting PBZT and PPTA fibers, which are based on rigid-rod polymers. The trick to impart conductivity is to mix metallophthalocyanine into the spinning dope (Wynne *et al.*, 1985; Polis *et al.*, 1989). However, the mechanical properties of these fibers deteriorate with increasing content of metallophthalocyanine.

There are some unusual applications that involve a combination of fibers. Consider the problem of static electricity that is generated from fabric-to-fabric or fabric-to-surface rubbing. It is generally thought of as a minor nuisance. It

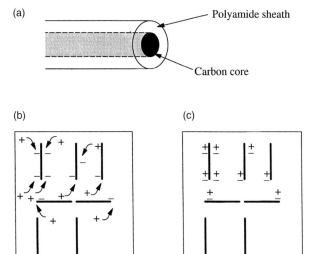

Figure 4.27 A blend of Nomex, Kevlar, and a core of pitch-based carbon fiber, P-140. (a) The carbon fiber forms the core. (b) Static charges on the polyamide fabric induce opposite charges on the carbon core. (c) When the induced charges on the carbon core build to a high energy level, surrounding air molecules ionize. Positive and negative ions neutralize charges on the polyamide fabric and the carbon core, thus dissipating the static electricity.

can, however, become a great danger in explosive atmospheres. It would be highly desirable to provide anti-static characteristics to a fabric made of Nomex fiber. To that end, Du Pont company produced a blended fiber called Nomex Delta A, which is a blend of Nomex, Kevlar, and a core of pitch-based carbon fiber, P-140. The carbon fiber forms the core (Fig. 4.27a). Static charges on the polyamide fabric induce opposite charges on the carbon core (Fig. 4.27b). When the induced charges on the carbon core build to a high energy level, surrounding air molecules ionize. Positive and negative ions neutralize charges on the polyamide fabric and the carbon core, thus dissipating the static electricity (Fig. 4.27c).

4.6.3 Flame retardant polymeric fibers

There is a great deal of interest in flame retardant textiles. This is understandable in view of the fact that many investigations show textiles to be the primary cause of flame spreading after ignition. The characteristics that provide high strength and stiffness also lead to high-temperature stability and flame resistance. Table 4.4 provides a comparison of high-temperature properties of some synthetic organic fibers (Irwin, 1997). The m-aramid and p-aramid in this table refer to Nomex and Kevlar type aramid fibers, respectively, while PTFE refers to poly tetrafluorethylene fiber. The limiting oxygen index (LOI) is a convenient index for rating the various fibers in regard to their flame resistance. LOI is the percentage of oxygen needed in an air atmosphere, i.e. oxygen/nitrogen atmosphere, to sustain burning. Recall that air contains about 21% oxygen; materials with LOI over 21 are not readily flammable in air.

Two basic approaches to deal with the problem of flammability are:

Table 4.4 *High-temperature properties of some synthetic organic fibers (after Irwin, 1997).*

	Percentage strength retained at 260°C	Time (h) for 50% strength loss at 180°C in air	Melting point (°C)	Shrinkage (%) at 180°C in air	Flammability LOI
Nylon 66	melts	100	254	4–11	24 (melts)
PET	melts	300	256	3–12	23 (melts)
PTFE	6%	years	310	7	95
m-aramid	58	years	410	0.4	29
p-aramid	63	years	none	<0.1	29

(i) Use a polymer whose chemical structure is flame retardant. A good example of such a fiber is the Nomex aramid fiber which is widely used as a flame retardant fiber.

(ii) Add a flame retardant additive to the flammable polymer. For example, a flame retardant conventional rayon fiber called the Lensing FR fiber has a thiosulfate as a flame retardant.

For more detail on this subject, the reader is referred to Hall and Horrocks (1993) and Holme (1994).

4.6.4   Hollow fibers

Hollow membrane fibers are required for many medical application, e.g. for disposable dialysis. Such fibers are made by using an appropriate fiber spinning technique with a special inlet in the center of the spinneret through which the fiber core forming medium (liquid or gas) is injected. The membrane material may be made by melt-spinning, chemical activated spinning or phase separation. The thin wall (15–500 μm thickness) acts as a semi-permeable membrane. Commonly, such fibers are made of cellulose-based membrane materials such as cellulose nitrate, or polyacrylonitrile, polymethylmethacrylate, polyamide and polypropylene (van Stone, 1985).

4.7   Applications of synthetic polymeric fibers

Synthetic polymeric fibers find extensive applications in all walks of life.
    Fibers such as nylon and polyester find applications in a wide variety of fields:

garments, hosiery, lingerie and underwear, raincoats, and a variety of sports apparel; home furnishings such as linen, carpets, upholstery, curtains, etc.; and industrial uses such as ropes, nets, conveyer and seat belts, tire cords, blend papers, tents, etc. Here we should mention an important item; wash and wear clothing, which has, over the years, become quite popular. Fabrics for such use generally consist of blends of natural and synthetic fibers, e.g. 60–70% polyester fiber and 40–30% cotton fiber. Cotton fiber, as described in Chapter 3, is about 95% cellulose. It is hydrophilic, i.e. it absorbs moisture readily by polar attraction. Water then acts as a plasticizer and thus makes wrinkling of a 100% cotton fabric easy. Polyester (poly ethylene terephthalate, a thermoplastic), on the other hand, is hydrophobic, i.e. it does not absorb water easily, stays unplasticized and thus wrinkle-free. In a humid atmosphere, however, the human body will feel sticky because of the inability of the polyester fabric to absorb moisture. A fabric made of a blend of cotton and polyester will provide a good combination of wrinkle or crease resistance as well as wear comfort in a humid atmosphere, i.e. a less sticky feeling.

Microfiber or microdenier fiber has become important in the fashion industry. The reason for this is the fine size of microfiber. It is finer than any natural textile fiber such as silk, wool, etc. Microfibers are nothing but polyester fibers with diameter $<5\,\mu m$ (or less than 1 den). Such fineness allows more filaments to be packed in the yarn. Fabrics using such tightly bundled filaments are said to have a 'buttery' texture like velour or washed silk. They also have the desirable characteristics of polyester, namely wrinkle-resistance and durability. Such fabrics are used in men's wear, women's wear, rainwear, active wear and home furnishings.

Polyolefins such as polyethylene and polypropylene are lightweight, hydrophobic and highly crystalline. It is this non-wetting characteristic of olefins that allows a sportswear made of, say, polypropylene, to wick moisture away from the wearer's skin. Other applications of polypropylene fiber include ropes, cords, fish nets, filters, as pile and backing in carpets, knitwear in the form of yarns blended with wool or cotton, and geotextiles. Polypropylene fiber (chopped) is also used for reinforcement of cementitius materials. It is worth pointing out here that the amount of fibers used in a reinforced cement or concrete product is generally very small (between 0.5 and 10 vol%) *vis-à-vis* a fiber reinforced polymer (over 50 vol%). The objective of introducing such fibers in cements and concretes is to improve their tensile and impact strength and to provide post-cracking ductility. Polyolefins are used extensively for upholstery, carpet, insulation and in geotextiles (see below). *Tyvek* is the trade name of a Du Pont fiber product made of spunbonded, nonwoven polyethylene fibers. It is highly tear resistant and is used for making envelopes, laboratory coats, etc. It is also used as insulation in housing construction. Tyvek can be used in the handling of radioactive materials. Fabrics made of this materials are tear-resistant, smooth, anti-static, and have a lint-free surface, qualities which allow it to be an effective barrier against particles as small as 0.6 μm. Thus, direct contact with radioactive material can be avoided. Polyethylene fibers of low molecular weight have limited applications because of low stiffness, poor creep resistance, and the inability to take dyes. Thus, they are mostly used for specialized nonapparel applications. Low density polyethylene (LDPE) fibers are used to make ropes and cords, filters, and protective clothing. High density polyethylene (HDPE) fibers are used in high altitude balloons, aerial tow targets, etc. Low density (*less than that of water!*) and resistance to water and rot have led to marine applications of polyethylene involving cordage that is useful for making ropes and nets that float in water. Fabrics for upholstery, tarpaulins, curtains, etc., are other uses of polyethylene. The development of the so-called ultra-high molecular weight polyethylene (UHMWPE) has opened up new applications such as lightweight body armor and as a reinforcement for polymers.

Another important application of thermoplastic fibers such as poly ether ether ketone (PEEK), Poly etherimide (PEI), and Vectran M and HS (Vectran is the trade mark of Hoechst liquid crystalline polymer) is in making thermoplastic matrix composites. Commingled yarns of the reinforcement and matrix such as quartz/PEEK, glass/PEI, Vectran HS/M are used to make the composites wherein the matrix yarn fuses to form the continuous phase of the composite.

Elastomeric fibers based on polyurethanes such as Spandex or Lycra are used extensively in garments and general athletic wear where high durability and resistance to flexure and abrasion are required. Another attractive feature of Spandex (or Lycra) is that it can be dyed easily and formed into a very fine yarn. Applications of Spandex fiber in apparel include a variety of undergarments, swimwear, waistbands, straps, outerwear, sports and leisure wear, etc. Elastomeric fibers can be used as bare fibers in a fabric or covered with other textiles to produce what are called covered and core-spun yarns. The covered yarn has one or two textile yarns interlaced in opposite directions around a highly strained Spandex fiber core. In the core-spun variety, a sheath of another fiber is put on Spandex fiber. An air jet is sometimes used to produce what is called an air-entangled filament around a stretched Lycra. The air-entangled filament forms random loops on the surface of Lycra. An air-entangled Lycra filament is easy to hook onto another yarn with which it is combined. It is easy to visualize the usefulness of such a feature in knit constructions that might need anti-slip characteristics. Elastomeric fibers can be used to make fabrics with predetermined extensibilities (between 20 and 350%). Their characteristics of elastic retraction, form fitting, and ability to provide support forces combined with an aesthetically pleasing appearance have made their use in leisure and sportswear a great success.

Aramid fibers and ultra high molecular weight polyethylene fibers (UHMWPE) are used in a variety of applications. For example, marine tow ropes, mooring cables, anti-ballistic clothing, fishing nets, sail cloth are prime markets for soft, thin, flexible organic synthetic fibers. Of special importance is the use of aramid and polyethylene fibers in vests for protection against light armor. Kevlar aramid fibers provide an impressive array of properties and applications. There is a variety of Kevlar fibers: K29, K 49, K68, K 119, K129, K149, K_{LT}, etc., the main ones being K 29 and K 49. Some applications include:

- Reinforcement for tires (belts or radial tires for cars and carcasses of radial tires for trucks) and, in general, for mechanical rubber goods.
- Ropes, cables, coated fabrics for inflatables, architectural fabrics, and for ballistic protection fabrics. Ship to shore mooring lines are made of aramid fibers because of its resistance to saltwater corrosion. Vests made of Kevlar 29 are used by law enforcement agencies in many countries. Later we discuss this application in some detail.
- Reinforcement of epoxy, polyester, and other resins for use in aerospace, marine, automotive and sports industries. We have previously mentioned the vibration damping capacity of Kevlar aramid fiber. Layers of woven Kevlar are used in skis for damping purposes and, of course, to reduce the weight. Kevlar is used as a protective sheath in fiber optic wave guides and to reinforce optical fiber cables because of its high tensile modulus and strength, and low electrical conductivity.
- Kevlar pulp can be used in thermosets, thermoplastics and elastomers. Injection-moldable pellets consisting of short Kevlar fibers in a thermoplastic resin can be made. These can be blended with unreinforced resins to obtain a composite with the desired volume fraction of reinforcement.

Nomex fiber has excellent fire resistance and is frequently used in applications requiring resistance to fire. Nomex and Kevlar fibers, individually or in a blend, are used to make fire protection garments. They are also used extensively in electrical insulation, thermal liners and moisture barriers.

Engineered fabrics are used for many industrial applications such as warehouses, swimming pool bubbles, etc. These fabrics, being very light, provide very simple air-supported structures. Figure 2.15 in Chapter 2 shows an example of such an air-supported structure. The walls should be light but strong enough to withstand snow, ice, sun and winds. Such structures can do without posts or pillars as they are supported by slight interior pressure. Frequently, the structure must be translucent to let light in.

Armor

Both polyethylene and aramid fibers have increasing applications in a variety of armor products. Body armor is a high performance system for protection against rifle bullets. Armor containing these fibers together with a polymeric matrix in the form a composite can provide ballistic protection as well as structural

Film

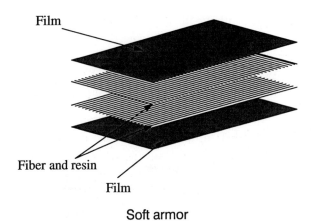

Fiber and resin

Film

Soft armor

Figure 4.28 Spectra Shield (a product of AlliedSignal) is a body armor that is made by means of woven fabric of Spectra polyethylene fiber. This figure shows a schematic of cross-plied (0°/90°) Spectra fibers in a resin matrix.

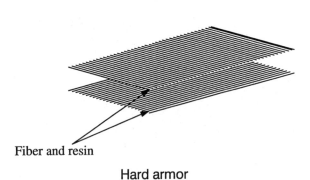

Fiber and resin

Hard armor

support. We give a brief description of this important application of such fibers.

Spectra Shield is a product of AlliedSignal that is made by means of a woven fabric of Spectra polyethylene fiber. Figure 4.28 shows a schematic of cross-plied (0°/90°) Spectra fibers in a resin matrix. Helmets, hard armor for vehicles, and soft body armor are shown in Fig. 4.29. The helmet manufacture involves a special version of Spectra Shield, a special shell design, and a three-way adjustable liner of shock absorbing foam padding. These helmets were used by the UN peacekeeping troops from France in 1993 and were introduced to police forces in the US and Europe. A soft body armor of polyethylene fiber consists of five Spectra Shield plies that slide into the pockets in the body armor.

Kevlar 29 and versions thereof (K 129 and K_{LT}) are also used extensively in lightweight body armor as well as composite liners (with vinylester, polyester or epoxy as the matrix). A quick look at the properties of different Kevlar aramid fibers in Table 4.2 shows why K29 is better than K49 for lightweight body armor applications. K29 has a higher strain to failure than K49. That means that the total work of fracture, i.e. the area under the stress–strain curve, is larger for K29 than K49. Hence, the energy absorbed in the fracture process is higher for K29

(a)

Figure 4.29 (a) A helmet reinforced with Kevlar aramid fiber and a bullet resistant vest made of Kevlar aramid fiber.

than K49. Additionally, there are some common features of Kevlar fibers. All of them show a very fibrillar fracture. Generally, a compact plain weave fabric structure is used to make protective armor. It is thought that slippage at the crossover points under impact contributes to the absorbed energy.

4.7.1 Polymeric fibers as geotextiles

Geotextiles have become one of the most important fields of application for synthetic polymeric fibers. In view of their great importance, we describe them in a separate section. Textiles made of synthetic polymer fibers are used in various applications to address a variety of solids-related problems in civil engineering such as soil support, stabilization, separation and filtration, reinforcement of

Figure 4.29   (b) Lightweight armor products made of Kevlar aramid fiber (courtesy of Du Pont).

embankments, stabilization of railroad track beds, highways, airports, waterways, erosion protection, etc. These materials are commonly made of fabrics of polyester, polypropylene and polyethylene fibers and have come to be known as 'geotextiles' simply because they are textiles used in the earth. Sometimes the term geosynthetics is used to underscore the fact that synthetic polymeric fibers are used. Polypropylene is by far the most common polymer used in geosynthetics. The American Society of Testing and Materials (ASTM) provides us with a convenient but rather broad definition of a geotextile as '. . . any possible technical material used with foundation, soil, rock, earth or any geotechnical related material as an integral part of a manmade project, structure or systems'. In a more general vein, we can define geotextiles as permeable fabrics that are used with soil to separate, filter, reinforce, protect or drain. Early on, jute was used as a geotextile material, but it is susceptible to degradation by micro-organisms in the soil. This led to the use of synthetic fibers as geotextiles.

A large variety of geotextiles is available depending on the type of polymer used, type of fiber construction, and the method of making the fabric. Among the fibrous materials commonly used for making geotextiles (in decreasing order of importance) are: polypropylene, polyester, polyamide and polyethylene. The following fiber forms are commonly used: monofilament, multifilament yarn, staple yarn, slit film. These fiber forms are then used to make fabrics. The fabrics may be woven, nonwoven or knitted. Nonwovens are commercially very important for making filters. The term geomembranes is used to denote impermeable geotextiles. They are used as liners for canals and ponds, and for containment of wastes and leachates that may be generated.

Geotextiles perform the functions mentioned above for long periods of time, over 50 to 100 years. The number of civil engineering projects all over the world using geotextiles, especially projects involving protection of the coast and the sea bed from erosion, has grown tremendously. Examples span from parking lots, recreational areas such as tennis courts, streets and highways to airport runways, tunnels and bridge decks. Among the airports, Singapore airport is a very good example where geotextiles are indispensable. This airport is built on a former lagoon and such a construction would not have been conceivable without extensive use of geotextiles. Geotextiles also form a critical component in any leachate collection/leak detection system in hazardous waste management facilities. In the United States, the Hazardous and Solid Waste Amendments of 1984 require all new hazardous waste storage and disposal facilities to be double lined with flexible membrane liners and there must be leak detection between the two liners.

A parameter called permittivity of geotextile (p) describes the ability of a geotextile to transmit water across the fabric. Permittivity p is given by the following relationship:

$$p = k_n/t$$

where k_n is the cross-plane permeability coefficient and t is the geotextile thickness at the normal pressure on the geotextile. The cross-plane permeability coefficient, k_n, is given by *Darcy's law*:

$$q = k_i A$$
$$= k_n (\Delta h/t) A$$

where q is the flow rate, Δh is the head loss across the geotextile and A is the area across which the flow occurs. Darcy's law, which forms the basis of the working of geotextiles, is really a general expression for flow through a porus medium, and understandably has applications in many areas.

There are many companies involved in the field of geotextiles. Among the important ones are Phillips Fibers, a subsidiary of Phillips Petroleum (Dutch/British) and Polyfelt of Austria. Petrofiber is a polypropylene fiber, made by Phillips Fibers, suitable for applications in pavement rehabilitation. This is a short length fiber and can be used by dispersing in the asphalt based matrix to improve the stability, ductility and fatigue life of the asphalt. It can also be used for placement in asphalt cement crack and joint fillers. Polypropylene fibers are inert, water insoluble and fairly stable with regard to moisture. They show good chemical,weather and abrasion resistance. It bonds well with asphalt bitumen. Petromat, another Phillips product, is a needle punched, nonwoven, polypropylene fabric that is heat set on one side.

Chapter 5

Metallic fibers

Metals in bulk form are quite common materials and extensively used in engineering and other applications. Metals can provide an excellent combination of mechanical and physical properties at a very reasonable cost. One of the important attributes of metals is their ability to undergo plastic deformation. This allows the use of plastic deformation as a means to process them into a variety of simple and complex shapes and forms, from airplane fuselages to huge oil and gas pipelines to commonplace aluminum soda pop cans and foil for household use. What is less well appreciated, however, is the fact that metals in the form of fibers or wire have also been in use for a long time. Examples of the use of metallic filaments include: tungsten filaments for lamps, copper and aluminum wire for electrical applications, steel wire for tire reinforcement, cables for use in suspension bridges, niobium-based filamentary superconductors, and, of course, strings for various musical instruments such as violin, piano, etc. Highly ductile metals such as gold and silver can be drawn into extremely thin filaments. Filaments of such noble metals have long been used as threads in making Indian women's traditional dress called the *saree*.

Let us first review some of the important characteristics of metals, in particular, the ones that allow metals and their alloys to be drawn into fine filaments, and then describe the processing, structure, properties and applications of some important metallic filaments. Readers already familiar with the basic attributes of metals may turn directly to Section 5.2.

5.1   General characteristics of metals

Metals are generally crystalline materials. Under very rapid cooling rates, say, greater than $10^6 \, \mathrm{K \, s^{-1}}$, one *can* also produce amorphous metals. Crystalline metals have the following three common crystal structures:

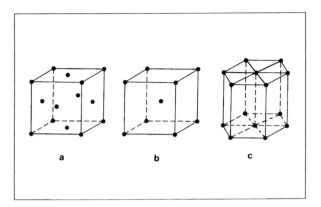

Figure 5.1 Typical structures of metals: (a) face centered cubic (FCC), (b) body centered cubic (BCC) and, (c) hexagonal close packed (HCP).

(a) face centered cubic (FCC);

(b) body centered cubic (BCC);

(c) hexagonal close packed (HCP).

Figure 5.1 shows these structures. Examples of FCC metals are aluminum, copper, gold, silver and γ-iron (austenite). They are all very ductile and can be drawn to less than 100 μm in diameter. This ability of the FCC and BCC metals to undergo large-scale plastic deformation stems from the large number of slip systems in these structures. BCC metals include α-iron (ferrite), tungsten, molybdenum and niobium. HCP metals include beryllium, magnesium, zinc and cadmium. They are not very ductile and it is not very easy to give them a filamentary form.

Metals are, of course, characterized by metallic bonding wherein we have a smeared cloud of electrons. This type of nonlocalized bonding results in exceptional mechanical, thermal, electrical, and magnetic characteristics. Metals can have a range of elastic modulus and strength values. For example, Young's modulus can range from a low of 17.5 GPa for lead to a high of 420 GPa for tungsten. The yield strength and ultimate tensile strength values also vary considerably, not only from metal to metal, but also for a given metal depending on the amount of mechanical working, alloying additions, and heat treatments used, which may involve a variety of phase transformations to modify the microstructure. High strength values can be obtained in filaments of tungsten, high carbon steels subjected to a heat treatment called patenting, and austenitic stainless steels after a large amount of plastic deformation.

In general, metals can be worked extensively, either at room temperature or at high temperatures. This is so mainly because of the availability of a large of number slip systems for plastic deformation. This allows us to use metal drawing techniques to obtain filamentary metals. Metallic fibers are, generally, not spun from a molten state, although this can be done in some cases (see Section 5.2). When metals are cold worked (i.e. below the recrystallization temperature), they

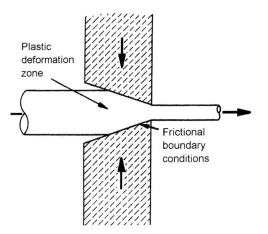

Plastic
deformation
zone

Frictional
boundary
conditions

Figure 5.2 Schematic of
wire drawing process for
metals. The process
exploits the ability of
metals to undergo large
scale plastic deformation.
The material is pulled
through a conical die that
forces a reduction in the
diameter. A variety of
shapes, round, hexagon,
square, etc. can be drawn.

tend to work harden. This phenomenon of work hardening, also called strain
hardening, results in an increase in the strength of the metal and a concomitant
decrease in its ductility. The strength increment has its origin in an increase in
the dislocation density in the metal: the strength increases as the square root of
the dislocation density. The cold working processes used to make wires and filaments invariably result in an increase in the dislocation density and, consequently, in the strength. It should be remembered that if such a strain hardened
metallic fiber is exposed to high temperatures (specifically, above the
recrystallization temperature of the metal), then the dislocation density will
decrease and the strength increment will be lost.

5.2   Processing of metallic filaments

Metals can be given a fibrous form by a variety of techniques. The most common
one is, of course, the technique of wire drawing. Wire drawing can be used to
obtain filaments down to 100 μm diameter. In order to obtain metallic filaments
smaller than this size, one must resort to unconventional techniques. We describe
below some of these techniques.

5.2.1   Wire drawing

Most metals with FCC and BCC structures can be easily drawn into a wire or
filamentary form. The process exploits the ability of metals to undergo large-
scale plastic deformation. A schematic of the wire drawing process is shown in
Fig. 5.2. The material is pulled through a conical die that forces a reduction in
the diameter. Lubrication must be used to decrease the frictional forces. The
entrance to the die is so shaped that the wire entering the die will also draw some

lubricant with it. A variety of cross-sectional shapes, round, hexagon, square, etc. can be drawn. The inner part of the die that comes in contact with the metal being drawn is made of hard material such as tungsten carbide or diamond. Almost all refractory metals are subjected to hot drawing, i.e. working at temperatures above their recrystallization temperature. This allows a higher ductility than is possible during cold working. Non-refractory metals are generally cold drawn, i.e. below their recrystallization temperature. It can be easily appreciated that proper lubrication is very important in all wire drawing. Graphite and molybdenum disulfide, both possessing a lamellar structure, are used as solid lubricants in the drawing of refractory metals such as tungsten and molybdenum. Tungsten and molybdenum are drawn at temperatures between 500 and 700°C with dry graphite film as a lubricant. In fact, colloidal graphite in a liquid carrier is used as lubricant. The carrier volatilizes when the wire is heated for drawing, leaving a solid film of graphite. At room temperature, use of graphite is not common because of the problems of removal. Molybdenum disulfide is used as a lubricant up to about 400°C, mainly because it starts oxidizing at about this temperature.

Work done in wire drawing

Mechanical work is done when a metal is drawn into a filament. When strains are large as in the mechanical working of metals, it is convenient to treat this in terms of true stress and true strain rather than engineering stress and engineering strain. True stress is load divided by the instantaneous cross-sectional area while true strain is the change in length divided by the instantaneous gage length. Engineering stress is the load divided by the original cross-sectional area while engineering strain is the change in length divided by the original gage length. Mechanical work done per unit volume, W, during deformation processing, can be written as

$$= \frac{1}{V}\int\sigma(\ell)A\,d\ell = \frac{1}{A\ell}\int\sigma(\ell)\,A\,d\ell = \int\sigma(\ell)d(\ln\ell) = \int\sigma d\epsilon \qquad (5.1)$$

where V is the volume, A is the cross-sectional area, ℓ is the length, σ is the true stress and ϵ is the true strain. As can be seen from Eq. (5.1), the work done per unit volume, W, is nothing but the area under the true stress–true strain curve. If we restrict ourselves to plastic deformation between two strain values, say ϵ_1 and ϵ_2, then the integral in Eq. (5.1) will be evaluated between these limits of strain. Another simplification commonly used is to replace $\sigma(\epsilon)$ by the mean value of flow stress between the strain limits of ϵ_1 and ϵ_2. Thus, Eq. (5.1) becomes

$$W = \bar{\sigma}(\epsilon_1 - \epsilon_2) \qquad (5.2)$$

where $\bar{\sigma}$ is the average flow stress $(\sigma_1 + \sigma_2)/2$. This simplification of taking an average value of flow stress is valid in hot working or in severely cold worked metals. The reader should realize that in plastic working, one aims at avoiding fracture of the metal, i.e. at all times the maximum tensile stress pulling the material through the die should be less than the strength of the drawn material coming out of the die. Any friction will increase the drawing force. Frequent breaking of wire is, of course, highly undesirable from a productivity point of view, because a wire break means the wire must be pointed again, rethreaded, and the process restarted. This point is discussed further in Section 5.3.

Wire drawing can be carried out cold or hot. Although hot drawing is common for refractory metals, cold drawing is more commonly done with non-refractory metals such as steel, copper, gold, silver, etc. Friction, as well as the large amount of plastic deformation, can result in a rise of wire temperature of several hundred degrees celsius. We can take into account the frictional and other non-ideal work during drawing by writing the total work done in wire drawing as,

$$W_t = W_i + W_f + W_r \qquad (5.3)$$

where W_i is the ideal work, W_f is the frictional work, and W_r is called the redundant work. Redundant work is the work that is done in excess of that required to produce the desired shape. In practice, it is very difficult to separate the frictional and redundant components of work done. One therefore defines a drawing efficiency parameter, η, as follows:

$$\eta = W_i / W_t$$

This drawing efficiency is a function of the die angle, amount of deformation produced in a pass, lubrication, etc. In general, the value of η varies between 0.5 and 0.65 (Hosford and Caddell, 1983). In terms of an average stress, one can write for the total work done

$$W_t = \frac{1}{\eta} \int \sigma d\epsilon = \frac{W_i}{\eta} \qquad (5.4)$$

For many metals, the flow curve, i.e. the true stress versus true plastic strain curve, can be expressed as a power law,

$$\sigma = K\epsilon^n \qquad (5.5)$$

where n is called the strain hardening exponent and K is the strength coefficient. It can be shown that the strain hardening exponent equals the uniform strain (Meyers and Chawla, 1984). Generally, the strain hardening exponent has an upper limit of about 0.5, i.e. the maximum reduction in area will be less than 50%. In practice, area reductions are less than 30% and for most metals, $0.1 < n$

< 0.5. A log–log plot of Eq. (5.5) gives slope $=n$ and $K=\sigma$ at $\epsilon=1$. Note that the strain hardening rate is not the same as the strain hardening exponent. In fact, we can write from Eq. (5.5)

$$n=\frac{d(\log \sigma)}{d(\log \epsilon)}=\frac{d(\ln \sigma)}{d(\ln \epsilon)}=\frac{d\sigma}{\sigma}\cdot\frac{\epsilon}{d\epsilon}$$

or the strain hardening rate is

$$\frac{d\sigma}{d\epsilon}=n\frac{\sigma}{\epsilon}$$

For a material obeying power law work hardening (Eq. (5.5)), it can be shown that

$$W_t=\left(\frac{1}{\eta}\right)\frac{K\epsilon^{(n+1)}}{n+1}$$

If work hardening is small (e.g. during hot working), one can use an average stress, $\bar{\sigma}$, and obtain the following expression for W_t:

$$W_t=\frac{1}{\eta}\int\bar{\sigma}d\epsilon=\frac{\bar{\sigma}(\Delta\epsilon)}{\eta}$$

where $\bar{\sigma}$ is the average flow stress over the range $\Delta\epsilon$. Equation (5.7) is essentially the same as Eq. (5.2).

5.2.2 Taylor process

Conventional wire drawing methods are quite reasonable for producing wires of Al, Cu, Ti, W, Ta, Mo, steels, etc. with diameters down to 100 μm. Production costs increase tremendously below this diameter. Donald (1987) defines a metallic wire less than 100 μm in diameter as *microwire*. Metallic wires of diameters down to 10 μm or less can be obtained by the Taylor process, so called after the person who first used it to produce a variety of fine metallic wires (Taylor, 1924). For a detailed review of developments in this process in the former USSR, USA, Europe and Japan, see Donald (1987). The basic process is indicated in the schematic shown in Fig. 5.3. A thick metallic wire is encased in a sheath of a sacrificial material (e.g. glass), heated to a temperature where the sheath becomes quite soft and the core wire melts or softens. This is followed by drawing in a plastic state down to very fine diameter and removing the sheath material by etching. Production of fine metallic wires is very expensive due to the cost of the Taylor wire drawing process. It turns out, for example, that for steel wires of diameters <25 μm, the cost of wire production becomes constant per unit length and not per unit weight, i.e. the material cost at such fine diameters is not very high but the processing costs are. The following requirements

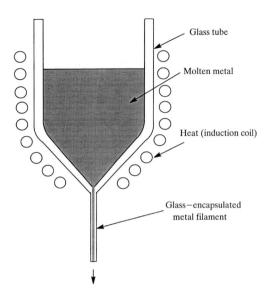

Figure 5.3   Schematic of the Taylor process.

must be met in order to produce fine metallic wires by the Taylor method (Donald, 1987):

(a) The glass must not react with metal at the drawing temperature.

(b) The working temperature of glass must be greater than the melting point of the metal and below the temperature at which the vapor pressure of the metal is too high.

(c) The viscosity of glass in the processing temperature range should be such that easy fiber drawing occurs.

(d) The coefficient of thermal expansion of the glass should be equal to or slightly less than that of the metal. If this condition is not met, then high enough thermal stresses can be generated in the glass to break it and a continuous fiber will not be obtained.

(e) The glass must become highly viscous before the metal becomes solidified. If not, the metallic core will become solid while the glass sheath continues to draw, resulting in fracture of the metal.

In practice, the restrictions enumerated above limit the sheath glass material to one of the low thermal expansion, borosilicate (Pyrex-type) glasses or fused silica.

5.2.3   Metallic fibers by spinning

Spinning fibers from a melt is perhaps the most common technique used to make fibers from polymeric and silica-based glasses. Such a technique, however, is not very suitable for metallic fibers because molten metals have very low viscosity, rather like that of water, and rather high surface energy. These characteristics of metals generally preclude the use of casting or extrusion from a molten state to make metallic fibers. If one were to try such a technique, what would happen is that as soon as the filamentary jet comes out of the die or spinneret, it breaks up into droplets because of a fluid instability phenomenon due to surface waves called Rayleigh waves. The amplitude of Rayleigh waves increases exponentially with the distance from the orifice, making the molten metal jet unstable and breaking it up into droplets. The *sine qua non* condition for spinning a fiber from the liquid state is that the free-liquid jet must be made stable over a length sufficient to allow it to freeze before breaking into droplets. This kind of jet stabilization technique has been used to produce some crystalline ceramic fibers (see Chapter 6). It would thus appear that in order to make metallic fibers by melt spinning, one must resort to some technique that stabilizes the molten jet as soon as it comes out of the orifice. Such a technique has also been referred to in the literature as 'free flight melt-spinning into a gaseous environment'. Pond (1961) ejected the molten metal through a nozzle into an inert atmosphere at a speed high enough to prevent break up into droplets before solidification. The use of helium as a high heat transfer coefficient cooling medium promotes rapid solidification of the melt and helps in maintaining a stable jet. Engelke (1967) ejected molten metal through an orifice into an appropriate liquid that surrounds and flows with the molten metal stream. Alternatively, one can use a method due to Alber and Smith (1965) which involves the use of an alloy that forms a stable oxide that is insoluble in the molten metal. The molten alloy is ejected into an oxidizing atmosphere and an oxide film forms on the molten jet. Lead has been made into filaments by this technique (Kikuchi and Shoji, 1989). The advantages with lead are that it has a low melting point, which allows a variety of spinneret materials to be used, and it tends to oxidize rapidly in air. The rapid formation of a lead oxide layer on the surface tends to stabilize the jet of molten lead against breaking into droplets. The important requirement in all fiber spinning processes is that a stable (i.e. laminar rather than turbulent) flow of liquid be maintained.

5.3   Microstructure and properties of metallic fibers

Metals when formed into wires can show rather high strength levels. The high strength stems from the phenomenon of work hardening that metals undergo during mechanical working. Work hardening, or strain hardening as it is sometimes called, refers to the fact that metals become stronger when they are cold worked (Meyers and Chawla, 1984), i.e. their strength or flow stress increases as a function of strain and correspondingly their ductility or toughness decreases. The phenomenon of work hardening has its origin in the increase in dislocation density produced during cold working of the metal. The modulus, however, does not change significantly with deformation. Another great advantage of metallic filaments is that they show very consistent strength values *vis-à-vis* ceramic fibers. The Weibull modulus of metallic fibers is an order of magnitude higher than that of ceramic fibers.

Beryllium, steel and tungsten can show good combinations of modulus, strength, and refractoriness. Beryllium, in particular, has a high modulus of about 300 GPa and an extremely low density 1.8 g cm^{-3}. Its toxic nature requires special handling, which makes it very expensive. Its strength is relatively low (~1300 MPa). We describe below the processing, microstructure, and properties of some important metallic filaments; tungsten, niobium-based superconducting filaments and steel.

5.3.1   Tungsten

Tungsten wires were originally developed for electric lamps and that still constitutes the major use of tungsten. It has a high melting point (3400°C), high modulus (414 GPa), and a very high density (19.3 g cm^{-3}). Tungsten has a BCC structure and has the somewhat unusual characteristic of being isotropic, even in the single crystal form. Besides the disadvantage of high density, tungsten oxidizes easily and the oxide of tungsten is likely to volatilize at high service temperatures. The advantages of tungsten include a high melting point, high elastic modulus and strength coupled with a high electrical resistivity.

Conceptually, and with a little bit of hindsight, the modern electric lamp is a very simple device: an electrically conducting filament enclosed in an evacuated glass enclosure. The filament emits light when heated to incandescence by the passage of an electric current. In 1879, Edison used a carbon filament for this purpose. Carbon was not very successful, it was replaced by osmium and then tantalum before tungsten filament was used. The success of tungsten filament is credited to some pioneering work done by Coolidge at GE involving sintering and drawing of tungsten filament. The tungsten filament works in vacuum or surrounded by inert gases at about 2600°C. The high melting point of tungsten (3400°C) allows it to be operated at such a high temperature. Generally, the tungsten filament is wound in the form of a helical coil, see Fig. 5.4a. The coiled filament results in a reduced cooling effect and a compact lamp design. Figure 5.4b shows a higher magnification view of the tungsten filament shown in Fig. 5.4a. Note the rather rough surface markings produced during the drawing process. The characteristic fibrous microstructure of a highly drawn tungsten filament, as seen in TEM, is shown in Fig. 5.5.

When used as an electric lamp filament at high operating temperatures, the tungsten filament creeps under its own weight and most of this creep occurs by grain boundary sliding. In order to minimize this creep problem, an elongated, bamboo-type grain boundary structure as produced by wire drawing is desir-

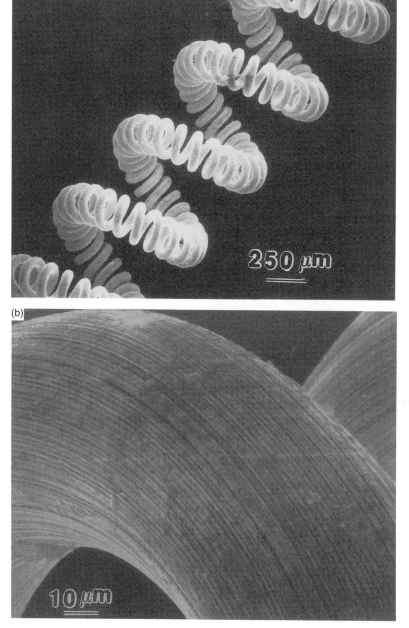

Figure 5.4 (a) Tungsten filament wound in the form of a helical coil. (b) The filament drawing process produces rather rough surface markings.

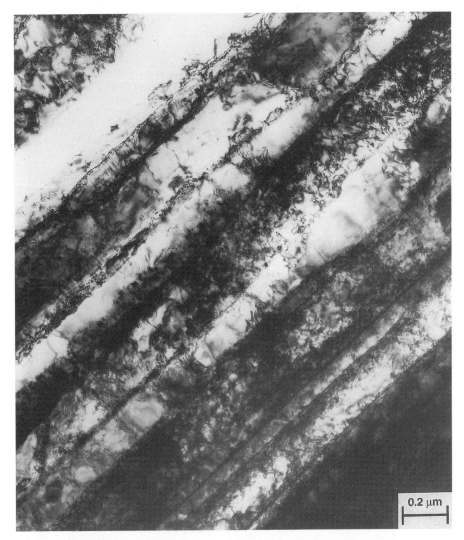

0.2 μm

Figure 5.5 Characteristic fibrous microstructure of a highly drawn tungsten filament. TEM (courtesy of B.P. Bewlay).

able. In such a bamboo-type grain structure, we have a small number of grain boundaries perpendicular to the filament axis. At one time, thoria (ThO_2) was used as an additive, primarily to control the grain growth in tungsten at high service temperatures and secondarily to give dispersion strengthening. Thoria (ThO_2) was added to tungsten as thorium nitrate prior to reduction of tungsten oxide by hydrogen. Thorium nitrate decomposed to form the oxide. The interstitial content, mainly oxygen and to a lesser extent nitrogen and carbon, can affect the ductility of tungsten wire. Very small amounts of oxygen, as little as fifty parts per million, are enough to embrittle tungsten. Although, thoria

(ThO$_2$) can be used as an alloying addition for tungsten, for incandescent electric lamp filaments, its use has largely been replaced by tungsten containing minor quantities of aluminum (Al), potassium (K) and silicon (Si). This type of tungsten filament is known as the AKS tungsten filament or *non-sag* tungsten filament (Wittenauer *et al.*, 1992). In fact, it is the potassium that is largely responsible for the controlled microstructure of tungsten filament as described below. Let us start from the very beginning. Tungsten ore is converted into ammonium paratungstate [5(NH$_4$)$_2$O.12 WO$_3$.11 H$_2$O], commonly referred to as APT. This APT is the starting material for making tungsten powder. APT is reduced to tungsten oxide (WO$_3$), which is doped with potassium disilicate and aluminum chloride (Bewlay *et al.*, 1991). Tungsten oxide is reduced to fine tungsten powder (<5 µm), pressed, and sintered into an ingot. Sintering of doped tungsten involves passage of an electric current ($\sim$5000 A) through the cold pressed ingot. A temperature of 3000°C is reached. At such high temperatures, impurities volatilize and rapid densification occurs and about 75 ppm of elemental potassium is left (Bewlay, 1991; Briant and Bewlay, 1995). Tungsten filament is made by drawing this tungsten ingot made via powder metallurgy. The ingot is rolled and/or swaged, followed by drawing down to a diameter less than 100 µm by using a series of dies. A 100 W bulb has a 50 µm diameter tungsten filament. The exact diameter of filament depends on the voltage and wattage of the lamp.

As noted earlier, creep in tungsten filament occurs by grain boundary sliding. Such a deformation makes the filament sag under its own weight and form a neck where it eventually breaks. Pure, recrystallized tungsten forms a bamboo-type grain structure, i.e. the grains occupy the entire diameter of the filament and the grain boundaries are aligned transverse to the filament axis. Under the service conditions for the tungsten filament, such boundaries will undergo sliding and lead to failure of the filament. The addition of potassium to tungsten results in an interlocking grain structure, which results in a reduced rate of grain boundary sliding and longer life for the filament than that of the undoped filament. The beneficial effect of potassium comes about because it helps to control the grain shape. After sintering the doped tungsten ingot has pores that contain elemental potassium. With wire drawing, these pores assume an elongated, tubular structure. When this material is annealed at a high temperature, these tubular structures containing potassium vapor become unstable as per Rayleigh waves on the surface of a cylindrical fluid (see Chapter 4). It is important that the deformation during drawing should be large enough to produce potassium cylinders with an aspect ratio >10, otherwise they will spheroidize to a form a single bubble (Briant, 1989; Vukcevich, 1991). Figure 5.6 shows an example of such bubbles in a transmission electron micrograph of a tungsten filament. The objective here is to have a high density of such tiny bubbles so that they pin the grain boundaries effectively against sliding, thus

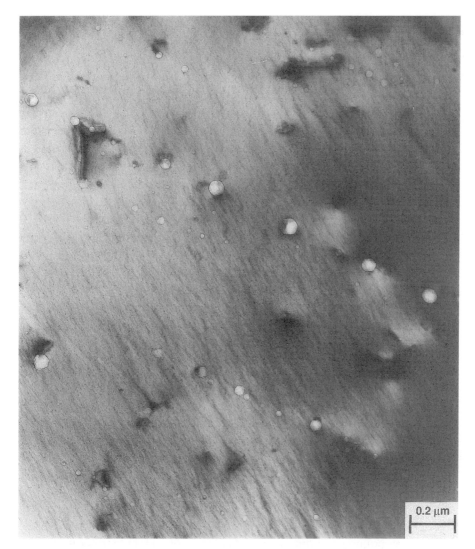

Figure 5.6 An example of potassium bubbles in a tungsten filament. TEM.
(courtesy of B.P. Bewlay).

resulting in a microstructure that is stable at the operating temperature. Non-sag tungsten filament can be regarded as a unique form of dispersion strengthened alloy in that it has elemental potassium as an alloying element in the form of tiny bubbles in the BCC tungsten lattice. These gas bubbles retard the recrystallization of the wire and give it a very superior creep resistance at the high temperatures prevailing in a glowing lamp and thus a longer life than that of the undoped filament. Figure 5.7 shows the superior creep resistance of the doped tungsten filament compared to that of an undoped one (Horascek, 1989).

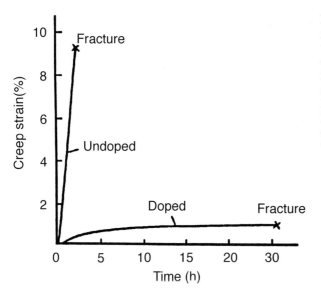

Figure 5.7 Creep curves for doped and undoped tungsten filaments (0.9 mm diameter) at 2800 K, under a tensile stress of 13 MPa (after Horascek, 1989). Note the lower creep strain in the doped tungsten *vis-à-vis* the undoped tungsten.

5.3.2 Niobium-based superconducting filaments

Niobium-based superconducting filaments such as Nb–Ti and Nb_3Sn, both embedded in a copper matrix, are used in making superconducting magnets. The alloy system Nb–Ti (about 50 wt%) is a typical, ductile metallic alloy (BCC structure) that can be extruded and drawn to very fine filaments when embedded in a copper matrix. One of its largest uses is in magnetic resonance imaging (MRI) systems (see Section 5.4).

A pertinent question here is why does one need the superconductor in a filamentary form? The answer lies in a phenomenon called *thermal runaway.* Dissipation of energy, associated with flux movement, can cause thermal runaway in a superconductor. Thermal runaway refers to any rapid build up of heat in the superconductor leading to a conversion of stored energy in the wire (recall the high current density that is used) into heat. In order to prevent such a catastrophe, the wire is designed such that outflow of heat to the surroundings is always greater than the rate of heat dissipation. This requirement means that the wire should have a very small diameter and should be arranged inside a high thermal conductivity material such as high purity copper. It also leads to another key requirement of these superconductors, namely that of magnetic flux pinning. This is achieved via proper metallurgical treatments involving a certain amount of cold working (extrusion and wire drawing) followed by an annealing treatment. Such a treatment results in a dislocation cell wall structure and precipitates (α-Ti), both of which act as flux pinning agents. A Nb–50% Ti alloy has a β microstructure with α-phase precipitates which are non-superconducting. An annealing treatment of 48 h at 375°C will result in about 11% of α-particles.

More α-particles will reduce the ductility of the alloy. An annealing temperature greater than 375°C will result in softening, fewer dislocations, and a lower J_c. Cold work after annealing leads to a refined structure and an increase in the J_c (Hillman, 1981). An important consideration during the drawing process is to avoid the formation of intermetallic defects from scratches (Hillman, 1981). A scratch on the surface of the NbTi rod can form a hard ball of oxidized NbTi about 1 μm in diameter. During extrusion, there is a snowball-like effect causing mechanical alloying and forming a NbTi+Cu mixture. Annealing treatment can cause the formation of $(NbTi)Cu_2$ and $(NbTi)_2Cu$ intermetallic compounds.

Another niobium-based superconductor is Nb_3Sn. It has an A-15 crystal structure and is a very brittle intermetallic, which is not easy to form. It has a $T_c=18$ K and $H_c>12$ T. It is extremely brittle, strain to fracture ~0.2%, that is, no plasticity. The way around this problem is to initially draw niobium wires in a bronze (Cu–Sn alloy) matrix. Niobium filaments embedded in a (Cu–13 wt% Sn) bronze matrix, with an R ratio=Cu–Sn/Nb>3, are drawn through a series of dies. When the desired diameter of wire is obtained, a heat treatment at 650–700°C is given to drive out the tin from the bronze matrix, it reacts with niobium to form Nb_3Sn leaving behind a copper matrix. Thermal stresses involved during a temperature change of ΔT ~1000 to 4.2 K put the Nb_3Sn filaments in compression (0.4–0.6% strain). The final product has a very fine grain structure (grain diameter <80 nm).

5.3.3   Steel

It is important to take into account the physical metallurgy of a steel in order to understand the general structure–property relationships in steels, and especially in a filamentary, eutectoid steel. Steel wire or filaments are obtained by first producing steel rod and then reducing its diameter to the desired size. The microstructure of this starting steel rod has a great influence on the final product. If the steel rod is cooled from the high finishing temperature of rolling by letting it stand in air, it will have a rather coarse pearlitic structure. A steel rod having a coarse pearlitic microstructure is not amenable to further cold drawing to a finer diameter; faster cooling is required for that. The hot rolled steel rod is descaled by mechanical and/or chemical methods, followed by cold drawing through a single die or a series of dies. Dry, powdered stearate is used for lubrication for coarse wires while for fine wires (<0.8 mm) a wet lubricant is used. Nonmetallic inclusions, particularly the nondeformable variety, decrease the ductility (this is discussed further later in this section).

In its simplest form, steel is an alloy of iron and carbon. In general, there are other alloying elements present. Steel wires can vary in composition enormously, from simple C–Mn steel to alloy steels containing more than ten alloy additions. Control of impurities (mostly nonmetallic inclusions), chemical segregation,

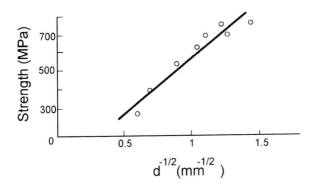

Figure 5.8 Inverse square root relationship between strength and diameter of a patented, pearlitic steel wire (after Embury and Fisher, 1966).

surface imperfections, chemical segregation, coupled with a uniformity in grain size can result in a very high quality product. The carbon content of steel wire may range from low to high, with the latter generally having carbon content between 0.5 and 0.9 wt%. A eutectoid steel has carbon between 0.8 and 0.9 wt% and consists of a 100% pearlitic microstructure. Steel in the form of wire can be very strong compared to the bulk steel although its modulus does not change very much. The strength of steel wire can span the range 0.8 to 5 GPa. Wires can range in diameter from coarse ($\geq$1.5 mm) to very fine ($\leq$0.1 mm). Low to medium carbon steels will have ferrite+pearlite microstructure. The term pearlite denotes a lamellar structure with the lamellae consisting of hard, orthorhombic cementite (Fe_3C) and soft, BCC ferrite (almost pure α-Fe). The high carbon (0.9% C) steel wires, commonly called piano wire and produced by the patenting process described later in this section, can have very high strength levels (~5 GPa), although the toughness levels will be rather low at such high strengths. It has a pearlitic structure, i.e. it has a lamellar structure with the lamellae consisting of hard, orthorhombic cementite (Fe_3C) and soft, BCC ferrite (almost pure α-Fe). The eutectoid decomposition of austenite (γ) phase into α and Fe_3C results in a structure consisting of a fine and uniform dispersion of cementite in ferrite. It is this characteristic structure that allows the wire to be drawn down to 75 μm and a very fine lamellar structure with interlamellar spacing between 50 and 100 nm is obtained. What is important to realize is that the key here is to understand the microstructure–property correlations. The strength of such a eutectoid wire increases as the inverse square root of the interlamellar spacing. The strength, σ, of pearlite is given by a relationship similar to the Hall–Petch relationship used for grain boundary strengthening of metals (Hall, 1951; Petch, 1953):

$$\sigma = \sigma_0 + \kappa \lambda^{-1/2}$$

where σ_0 is a friction stress term, k is a material constant and λ is the interlamellar spacing. It turns out that one can vary the interfiber spacing, λ, easily by controlling the cooling rate. In more practical terms, Fig. 5.8 shows an inverse

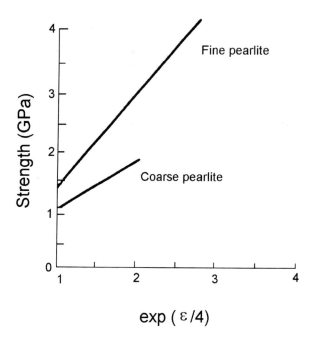

Figure 5.9 Effect of drawing or swaging operation on the strength of pearlitic steel wires. Both coarse and fine pearlite show a linear relationship between yield strength and exp (ϵ/4) where ϵ is the strain (after Embury and Fisher, 1966).

square root relationship between yield strength and diameter of a patented, pearlitic steel wire (Embury and Fisher, 1966). It can be shown that the inverse square root dependence of strength on the diameter translates into strength being proportional to exp (ϵ/4), where ϵ is the true plastic strain. We start by taking into account the fact that plastic deformation occurs at a constant volume, i.e., $A_0 l_0 = Al$, where A_0 and A are the original and instantaneous cross-sectional areas and l_0 and l are the original and instantaneous lengths, respectively. The true or logarithmic strain is given by:

$$\epsilon = \ln (l/l_0) = \ln (A_0/A)$$

If d and d_0 are, respectively, the instantaneous and original diameter of the wire or fiber, we can write

$$\epsilon = \ln (\pi d_0^2/4)/(\pi d^2/4) = 2 \ln d_0/d$$
$$\epsilon/2 = \ln d_0/d$$
$$d/d_0 = \exp (-\epsilon/2)$$
$$d = d_0 \exp (-\epsilon/2)$$
$$\sqrt{d} = \sqrt{d_0} \exp (-\epsilon/4)$$

From the Hall–Petch relationship, we know that $\sigma_f \propto (1/\sqrt{d})$, therefore $\sigma_f \propto (1/\sqrt{d_0}) \exp (\epsilon/4)$. Figure 5.9 shows the effect of the drawing or swaging operation on pearlitic steel wires. Both coarse and fine pearlite filaments show a linear relationship between strength and exp ($\epsilon/4$).

Processing and microstructure

The strength of steel wire is mainly controlled by its carbon content and its microstructure which depends on processing. The term microstructure in this case means ferritic grain size and the interlamellar spacing in the pearlitic regions. The microstructure is determined, just as in other steel products, by composition and final working conditions. In general, the yield strength and ultimate tensile strength increase with carbon content and concomitantly ductility and toughness decrease. In ferritic/pearlitic steels, the tensile strength of steel increases with:

(a) decreasing ferritic grain size;
(b) fine interlamellar spacing of pearlite;
(c) precipitation hardening;
(d) carbon content.

The effects (a) and (c) are generally combined by using finely distributed NbCN or VCN precipitates to give a fine grain size and precipitation hardening.

Most steel wire is produced by heavy cold drawing. The drawability of steel is a function of its microstructure, which is a very sensitive function of the cooling rate. For low carbon steels, the natural air cooling of the hot rolled steel results in a fine enough microstructure which enables cold drawing down to 95% reduction. For steels with carbon between 0.5 and 0.9 wt%, natural air cooling results in a coarse pearlitic structure which can be cold drawn in the 25–50% reduction range. In this carbon range (0.5–0.9 wt%), a process called *patenting* is used to produce steel wire. Very fine steel wires (0.1 mm or less in diameter) containing high carbon (0.9% C) can have very high strength levels (~5 GPa), although understandably the toughness levels will be low at such high strengths. Such a steel wire is known as piano wire. Its composition (in wt%) is 0.9 C, 0.37 Mn, 0.2 Si, 0.01 P, and 0.03 S. Its structure consists almost entirely of pearlite. Figure 5.10 shows a transmission electron micrograph of a fully pearlitic structure. The special processing, *patenting*, that is used to obtain this high strength wire involves drawing the wire down to about 500–750 μm, and holding at 900°C for a few minutes to transform completely to an FCC phase called austenite (γ phase). Austenite has much higher solubility for carbon, and takes all the carbon into solution. The wire is then cooled to 400–500°C in a molten lead or salt bath, which results in the eutectoid decomposition of austenite (γ) phase according to the following reaction

$$\gamma \rightarrow \alpha + Fe_3C$$

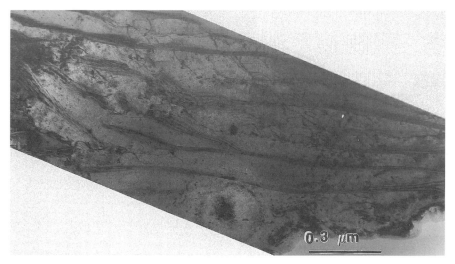

Figure 5.10   A fully pearlitic structure in a steel consisting of aligned lamellae of α-Fe (light) and Fe_3C (dark). TEM.

i.e. the γ phase decomposes into a very fine two-phase mixture, α and Fe_3C with a microstructure as shown in Fig. 5.10. The cooling rate is ~30°C/s. This structure, consisting of a fine and uniform dispersion of cementite in ferrite, allows the wire to be drawn down to 75 μm and a very fine lamellar structure. An interlamellar spacing between 50 and 100 nm is obtained. The lead patenting process is increasingly being replaced by fluidized bed patenting because of the health hazards associated with lead.

In more modern wire plants the whole patenting process has been eliminated. A controlled cooling system is used at the exit stands of hot-rolled rod at ~950–1100°C. The rod is cooled to a temperature between 750 and 900°C in a water box, followed by forced air cooling at a rate of 5–10°C s^{-1}. This treatment results in a microstructural fineness that is intermediate between that produced by air and lead patenting. Such a microstructure allows further reduction by cold drawing to 80%. Microalloy additions in high carbon rod can compensate for the lower tensile strengths achieved due to controlled cooling *vis-à-vis* the patenting route. Jaiswal and McIvor (1989) found the following relationship between the tensile strength, cooling rate and chemical composition:

$$\text{tensile strength (MPa)} = [267(\log CR) - 293] + 1029\ (\%C)$$
$$+ 152\ (\%Si) + 210\ (\%\ Mn) + 442\ (\%\ P)^{0.5}$$
$$+ 5244\ (\%\ N_f)$$

where CR designates the cooling rate in °C/s and the amounts of the different elements are in weight percentages with N_f being the free nitrogen.

Through a proper control of processing and microstructure of these steel

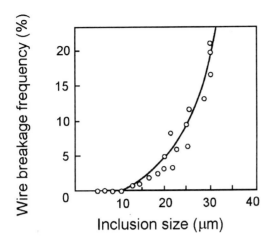

Figure 5.11 Effect of size of oxide inclusions on wire breakage (after Takahashi *et al.*, 1992).

wires, researchers at Nippon Steel (Takahashi *et al.*, 1992) have been able to produce steel wires with 2 GPa strength for use in suspension bridge cables. Another major application of steel filament is in tire reinforcement. Steel wire used for this purpose is commonly known as cord. It is instructive to understand the driving force behind the search for still higher strength steel wire for tire reinforcement. The desire to reduce the weight of an automobile was translated to a reduction in tire weight, which in turn led to efforts aimed at increasing the strength of steel reinforcement wire so that a thinner wire could be used for the same strengthening effect. This resulted in the addition of alloying elements in the composition of steel cord for tire reinforcement. Control of nonmetallic inclusions and a reduction in the degree of center segregation have resulted in a vast improvement in steel wire used for tire cord and bridge cables. Inclusions such as C, Mn, P, etc. tend to segregate to the center of the wire. It is typically made by drawing a 5.5 mm rod down to 0.15 mm diameter by a series of patenting steps. A key point in regard to the manufacture of steel cord is the number of breaks per ton of production in the drawing process (Takahashi *et al.*, 1992). Of course, the lower the number of such breaks the better. What is interesting is that the frequency of wire breaks is a function of the size of nonmetallic inclusions in the wire. Figure 5.11 shows the effect of size of oxide inclusions on wire breakage. Note that the wire break frequency tends to zero for inclusions of diameter 5 μm or less.

The strength of the steel wire is controlled by delaminations that appear along the wire axis when it is twisted during the cable strand formation. The genesis of nonmetallic inclusions in steel wire lies in the process of steel making. During solidification of steel, because of solute redistribution, elements such as C, Mn and P segregate between the dendrites. This inclusion containing molten steel ends up in the shrinkage cavities at the center of the bar, solidifies there and results in the centerline segregation. Figure 5.12 shows the segregation of manganese in

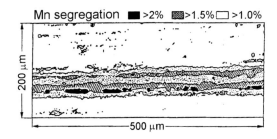

Figure 5.12 Segregation of manganese in the center of a rolled rod (after Takahashi *et al.*, 1992). Drawing limit strain as a function of the center segregation in the starting billet.

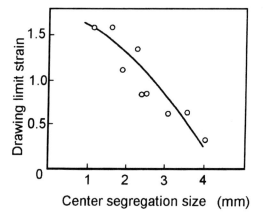

Figure 5.13 Drawing limit strain as a function of the center segregation size in the starting billet. A reduction in segregation size improves the drawability (after Takahashi *et al.*, 1992).

the center of a rolled rod. Any untransformed austenite in the central zone will transform to martensite (a hard, brittle phase) because of the presence of manganese there. Segregation of carbon can lead to cementite film formation at austenitic grain boundaries. Such segregation can severely limit the drawability of steel wire. Figure 5.13 shows the drawing limit strain as a function of the center segregation in the starting billet. A reduction in segregation size improves drawability. This tendency for centerline segregation can be minimized by finely dispersing the solute-enriched molten steel and pressure welding the shrinkage cavities where this enriched steel accumulates. This can be accomplished by lowering the pouring temperature of molten steel and electromagnetically stirring the molten steel (Takahashi *et al.*, 1992). After some extensive work at Nippon Steel, researchers came to the conclusion that the delamination cracking failure in steel wire can better be prevented by increasing the strength of the wire rather than increasing the total reduction in area during drawing. This can be done by increasing the carbon and silicon contents of steel and adding another element, chromium. Carbon and chromium add to the strength of the steel wire by decreasing the interlamellar spacing of pearlite while silicon simply gives solid solution strengthening. Figure 5.14 shows the effect of carbon and other alloying elements on the increase in the strength of a patented steel wire (Takahashi *et al.*, 1992). Now, silicon does not dissolve in cementite while chromium does. This

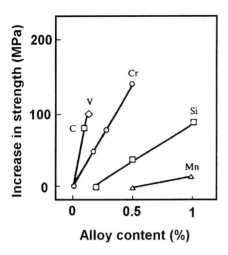

Figure 5.14 Effect of carbon and other alloying elements on the increase in the strength of a patented steel wire (after Takahashi *et al.*, 1992).

steel is pearlitic and the strength is governed by the interlamellar spacing. The steel wire must be subjected to heat treatment, drawing, and galvanizing operations. The galvanizing operation involves immersion of steel wires into a zinc bath at 450°C. This treatment would result in breaking up of the Fe_3C lamellae into small particles and the strength of the steel wire would be reduced. Takahashi *et al.* (1992) found that by increasing the amount of silicon in steel, a protective layer was formed around Fe_3C which prevented the break-up into small particles.

Mechanical properties of some commercial metallic wires are summarized in Table 5.1.

5.4 Applications

Although it is not generally appreciated, metallic filaments find a wide range of applications in modern day life. We provide below a summary of some important uses of metals in filamentary form.

Copper

Except for gold and silver, copper is the best metallic electrical conductor and copper wires are commonly used as electrical conductors. They are also used as magnet wires in electric motors, transformers, generators, electromagnets, etc. Copper and aluminum are commonly used for magnet wire because of their good electrical conductivity. Magnet wires are frequently coated by a variety of polymers such as nylon, aramid, polyester, polytetrafluoroethylene, polymide, etc.

Aluminum

Aluminum wires are used as electrical conductors. Although the electrical conductivity of aluminum not as good as that of copper, it is much lighter than copper and is therefore used as an electrical conductor. It should be pointed out that the electrical conductivity of aluminum is about 60% that of copper. This means that replacing a copper wire by aluminum requires a larger cross-section and thus a larger design of equipment. Sometimes a bimetallic wire consisting of aluminum and copper, made by coextrusion, is used as a conductor.

Steel

Steel in a wire or filamentary form is commonly used for reinforcement of tires and pressure hoses. It is also used for reinforcement of concrete, although more often than not, for this purpose it is in the form of a rod (diameter greater than 1 cm) rather than a wire. Steel wire with 0.8–0.9 wt % carbon (eutectoid steel) is used in pianos and in steel ropes used as cables in suspension bridges and for other structural purposes. We should emphasize the use of fine diameter steel

Table 5.1 *Typical properties of some commercial metallic wires.*[a]

Metal	Density ($g\,cm^{-3}$)	Melting point (°C)	Coefficient of thermal expansion ($10^{-6}\,K^{-1}$)	Young's modulus (GPa)	Tensile strength (MPa)
Al	2.7	660	24	70	300
Be	1.8	1350	12	310	1100
Cu	8.9	1083	16	125	450
W	19.3	3410	4.5	350	2890 (250 µm) 3150 (<125 µm) 3850 (< 25 µm
Mo	10.2	2625	6.0	330	2200
0.9% C Steel (0.1 mm dia.)	7.9	1300	11.8	210	4000
304 Stainless Steel (0.05 mm dia).	7.8	1535	9.0	198	2400

Note:
[a] These values are indicative only. Heat treatments (e.g. annealing, aging, etc.) and processing can alter the strength properties quite drastically.

wires in massive engineering structures such as single span suspension bridges which need steel cables whose strength ultimately depends on the strength of individual steel filaments composing the cables. Such cables can be one meter or more in diameter and contain 37 000 strands of steel filament.

An interesting application involves the use of very fine (15 μm) stainless steel filament as a conductive filler in a thermoplastic resin for electromagnetic shielding. According to Nippon Steel and Nippon Seisen Co., who developed this product called *Esbarrier*, 4–10 wt% of stainless steel fiber provides an effective shielding of 40–50 dB at 100 MHz.

Tungsten

Tungsten filaments are used in electric lamp bulbs. More than 90% of non-sag tungsten is used for incandescent lamps. A small amount is used as defrosting heating wires in automobile windshields. Another use is as heating elements for aluminum evaporation in coating applications. Tungsten fibers have been also used to reinforce copper and some nickel and cobalt based superalloys.

Lead

Heavy but soft materials such as lead reflect sound waves very efficiently. Because of its high density and low modulus, lead is an excellent acoustic insulator. Lead sheets are indeed used for this purpose, but not alone. This is because lead tends to creep easily under its own weight. A lead sheet is commonly laminated between say plywood or some other material. Toray Co. produced a lead fiber nonwoven fabric embedded in soft polyvinyl chloride with excellent sound insulating characteristics. Nonwoven lead fiber mat can be used as radiation shielding in nuclear installations. A blended product of fine lead short fiber with a resin has been used for X-ray shielding.

Niobium-based superconducting filaments

The superconducting magnet is the most important and expensive component of a magnetic resonance imaging (MRI) system. This is because flux densities above 0.5 T in a volume large enough for patient investigations can only be achieved with superconducting magnets (Oppelt and Grandke, 1993). Typical operating conditions for the superconducting magnets in an MRI apparatus would be a critical current density, $J_c > 1000$ A mm^{-2} at a temperature of 4.2 K and a magnetic field of 7 T. The solenoid coils consist of NbTi filaments embedded in a copper matrix.

Filamentary superconductors containing Nb_3Sn are used in applications such as high energy particle confinement in accelerators where very high magnetic fields are required. Such superconductors can show a $J_c = 2000$ A mm^{-2}, at 4.2 K and 10 T magnetic field.

Chapter 6

Ceramic fibers

In this chapter we provide a description of the processing, structure, and properties of high temperature ceramic fibers, excluding glass and carbon, which are dealt with in separate chapters because of their greater commercial importance. Before we do that, however, we review briefly some fundamental characteristics of ceramics (crystalline and noncrystalline). Once again, readers already familiar with this basic information may choose to go directly to Section 6.5.

6.1 Some important ceramics

We provide a summary of the characteristics of some important ceramic materials that have been converted into a fibrous form.

6.1.1 Bonding and crystalline structure

Ceramics are primarily compounds. Ceramics other than glasses generally have a crystalline structure, while silica-based glasses, a subclass of ceramic materials, are noncrystalline. In crystalline ceramic compounds, stoichiometry dictates the ratio of one element to another. Nonstoichiometric ceramic compounds, however, occur frequently. Some important ceramic materials are listed in Table 6.1. Physical and mechanical characteristics of some ceramic materials are given in Table 6.2. It should be noted that the values shown in Table 6.2 are more indicative than absolute.

In terms of bonding, ceramics have mostly ionic bonding and some covalent bonding. Ionic bonding means there occurs a transfer of electrons between atoms that make the compound. Generally, positively charged ions balance the negatively charged ions to give an electrically neutral compound, for example, NaCl. In covalent bonding, the electrons are shared between atoms. The characteristic high strength as well as brittleness of ceramic materials can be traced to these types of bonding which make the Peierls–Nabarro potential very high, i.e. inherent lattice resistance to dislocation motion is very high. Also, the number of slip systems available in ceramics is less than that in metals. Thus, unlike metals, a stress concentration at a crack tip in a crystalline ceramic cannot be relieved by plastic deformation, at least not at low and moderate temperatures. This has led to attempts at toughening ceramics by means other than large scale dislocation motion, for example, by incorporating fibers or second phases (Chawla, 1993).

6.1.2 Crystalline ceramics

Generally, metallic cations are smaller than nonmetallic anions. Thus, in crystalline ceramics, the metallic cations occupy interstitial positions in an array of nonmetallic ions. Common crystal structures in ceramics, shown in Fig. 6.1, are:

Table 6.1 *Some important ceramic materials.*

Single oxides	Alumina (Al_2O_3)
	Zirconia (ZrO_2)
	Titania (TiO_2)
	Magnesium oxide (MgO)
	Silica (SiO_2)
Mixed oxides	Mullite ($3Al_2O_3 \cdot 2SiO_2$)
	Spinel ($MgO \cdot Al_2O_3$)
Carbides	Silicon carbide (SiC)
	Boron carbide (B_4C)
	Titanium carbide (TiC)
Nitrides	Boron nitride (BN)
	Silicon nitride (Si_3N_4)
Intermetallics	Nickel aluminide (NiAl, Ni_3Al)
	Titanium aluminide (TiAl, Ti_3Al)
	Molybdenum disilicide ($MoSi_2$)
Elemental	Carbon (C)
	Boron (B)

(a)

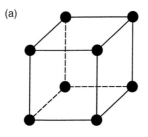

Figure 6.1 Common crystal structures in ceramics: (a) Simple cubic (or CsCl) structure ; (b) close-packed cubic (or NaCl), a variant of the face centered cubic (FCC) structure; (c) hexagonal close-packed (HCP).

(b)

(c)

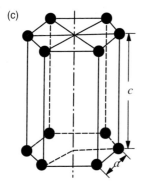

Table 6.2 *Physical and mechanical characteristics of some ceramic materials.*

Material	Density, ρ (g/cm^3)	Melting point (°C)	Young's modulus, E (GPa)	Coefficient of thermal expansion, α (10^{-6}/K)	Fracture toughness, K_{ic} (MPa m$^{1/2}$)
Al$_2$O$_3$	3.9	2050	380	8.5	2–4
SiC	3.2	—	420	4.5	2.2–3.4
Si$_3$N$_4$	3.1	—	310	3.1	2.5–3.5
MgO	3.6	2850	210	3.6	—
Mullite	3.2	1850	140	5.3	3.5–3.9

(1) *Simple cubic*. This is also called the cesium chloride structure. Examples are CsCl, CsBr and CsI. It is not as common as the other structures listed below.

(2) *Close packed cubic*. This is also called the NaCl structure and is really a variant of the face centered cubic (FCC) structure. Examples of this structure include CaO, MgO, MnO, NiO, FeO and BaO. Oxygen ions occupy the FCC positions while the metal ions occupy the interstices.

(3) *Hexagonal close packed*. Examples of this structure include ZnS and Al_2O_3.

Next we give a summary of important characteristics of some ceramic materials that are available in a fibrous form.

Alpha-alumina (α-Al_2O_3) is the thermodynamically stable phase of alumina. It has a hexagonal structure with aluminum ions at the octahedral interstitial sites. Each aluminum ion is surrounded by six equidistant oxygen ions. Figure 6.2a shows the hexagonal close packed structure of α-alumina. The A and B layers contain oxygen ions while the C layers contain aluminum ions and vacant sites. The C layers, indicated as C_1 and C_2, are two-thirds aluminum ions and one-third vacancies. This makes for charge neutrality. Figure 6.2b shows two slip systems, basal and prismatic, in a hexagonal structure. In alumina, at temperatures below 1000°C, slip can occur on these two slip planes. A fine-grained alumina structure is generally desirable because both low temperature strength and toughness increase with a decreasing grain size. For most crystalline ceramics, at ambient temperatures, the finer the grain size, the smaller will be the flaw size, and the higher the strength. At high temperatures, where creep becomes important, the strength of ceramics is mainly determined by the characteristics of the glassy phase that is frequently present at the grain boundaries. Hence, a larger grain size (i.e. lower grain boundary area) is desirable for high temperature creep resistance. Typical properties of monolithic alumina are given in Table 6.2.

Mullite is another important crystalline ceramic that is an oxide. It is a solid solution of alumina and silica in the compositional range 71–75 wt % alumina. Mullite is represented by the formula, $3Al_2O_3.2SiO_2$. It has excellent strength and creep resistance as well as low thermal expansion and conductivity. For more detailed information on the structure and properties of mullite the reader is referred to Schneider *et al.* (1994). Table 6.2 provides a summary of the properties of mullite.

Among nonoxides, silicon carbide and silicon nitride are two very important ceramics. Both are very hard and abrasive materials, and show excellent resistance to erosion and chemical attack in reducing environments. In oxidizing environments, any free silicon present in a silicon carbide or silicon nitride compact will be oxidized readily. Silicon carbide itself can also be oxidized at very high temperatures, the exact temperatures being a function of purity and

(a)

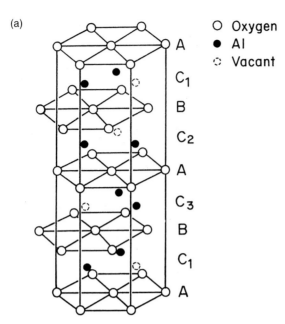

○ Oxygen
● Al
○ Vacant

A
C_1
B
C_2
A
C_3
B
C_1
A

Figure 6.2 (a) The hexagonal close-packed structure of α-alumina. (b) Two important slip systems, basal (0001) [0001] and prismatic $(01\bar{1}0)[2\bar{1}\bar{1}0]$, in a hexagonal structure.

(b)

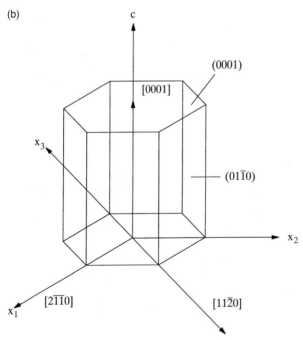

dwell time. Silicon carbide has two crystalline polymorphs: α-SiC (hexagonal) and β-SiC (cubic). The hexagonal α-silicon carbide has many polytypes which differ in the stacking sequence. Typical physical and mechanical properties of silicon carbide are given in Table 6.2.

There are two forms of silicon nitride, α and β, and both are hexagonal, with the c-axis of α-Si$_3$N$_4$ being about twice that of β-Si$_3$N$_4$. Contamination of Si$_3$N$_4$ with oxygen is a perennial problem. Typical physical and mechanical properties of silicon nitride are given in Table 6.2.

Boron and nitrogen can form the following BN compounds which are iso-structural polymorphs of carbon.

- α-BN. This has a hexagonal, layered structure similar to graphite with a theoretical density of $3.27\,\mathrm{g\,cm^{-3}}$.
- β-BN. This variety has a diamond cubic structure and is extremely hard like diamond. Its theoretical density is $3.48\,\mathrm{g\,cm^{-3}}$.
- γ-BN. This is also hexagonal, but with a higher density of $3.48\,\mathrm{g\,cm^{-3}}$.

The hexagonal variety of boron nitride shows natural lubricity because of easy cleavage along the basal planes. Boron nitride is applied generally as a paint, paste, or aerosol. Because of its structural similarity with lamellar graphite, hexagonal BN is sometimes referred to as white graphite. Graphite, however, is a good electrical conductor while BN is an insulator. Boron nitride shows a better resistance to oxidation than graphite.

Finally, an important form of boron nitride should be mentioned, pyrolytic boron nitride. It is manufactured by reacting ammonia and a boron halogenide at about 2000°C and depositing the BN vapor on a graphite substrate or mandrel. The characteristic feature of pyrolytic boron nitride is the high degree of crystal orientation with the hexagonal basal plane parallel to the mold surface and the c-direction perpendicular to the substrate.

6.1.3   Noncrystalline ceramics

Noncrystalline or amorphous (i.e. *without form*) ceramics are supercooled liquids. Liquids flow under their own mass, but they can become very viscous at low temperatures. Very viscous liquids (for example, honey in the winter time) have solid-like behavior although they maintain a disordered structure characteristic of a liquid, i.e. they do not undergo a transformation to a crystalline structure. Thus, noncrystalline ceramics, i.e. glasses, may behave, in many respects, like solids but structurally they are liquids.

When a liquid is cooled, molecular or atomic rearrangement occurs leading to a closer packing of molecules or atoms. We described the behavior of a liquid on cooling in Chapter 3. In particular, Fig. 3.2 in Chapter 3 shows a plot of specific volume (i.e. volume per unit mass) versus temperature. In the case of a crystalline material, at the melting point, T_m there occurs a precipitous drop in the specific volume and, correspondingly, a change in a variety of other properties. In the case of glassy materials, a sharp melting point is not observed. Instead, there occurs a gradual change in the slope of the cooling curve. This change in slope allows us to define a glass transition temperature, T_g, for glassy materials. At temperatures below the glass transition temperature, T_g, the supercooled glass becomes rigid and no further rearrangement of atoms or molecules occurs. Thus, glass, a noncrystalline solid, has a frozen-in structure of a liquid. However, unlike true liquids but like solids, it shows resistance to shear forces and shows a Hookean behavior, i.e. strain produced is linearly proportional to the applied stress. Silica-based glass fibers are described in Chapter 7.

6.2   Creep in ceramics

Since ceramics are used at very high temperatures, their behavior at high temperatures is very important. Inasmuch as ceramic fibers are also meant mainly for high temperature applications, it is important to examine their creep behavior. In most ceramic materials, the number of operative slip systems is limited because of their ionic/covalent bonding. As a result, they show low ductility or strain to fracture at low temperatures. At high temperatures, however, ceramics can show time-dependent phenomena, i.e. creep. A typical creep curve is shown in Fig. 6.3. The curve is divided into three stages: primary (stage I), secondary or steady state (stage II), and tertiary (stage III) in which accelerated damage leads to fracture.

In polycrystalline ceramics, creep can occur by cavitation and grain boundary sliding. This is commonly made worse by the presence of any glassy phase at the boundaries. Commonly, under creep conditions, cavitation is observed in the intergranular glassy phase. At low stresses and high temperatures *diffusional creep* can occur. The mechanism responsible for this is the operation of vacancy currents. At high stress levels, *power law creep* occurs. As the creep mechanism changes from diffusional to power law, the damage rate increases significantly. At a homologous temperature (the ratio of temperature in kelvin to the melting

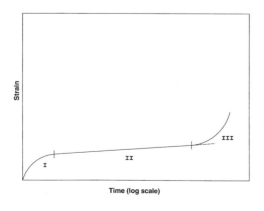

Figure 6.3   A typical creep curve.

point in kelvin, T/T_m) equal to or greater than 0.6, creep in ceramics can occur by dislocation climb.

Steady state creep rate ($\dot{\epsilon}$) (the creep rate in stage II in Fig. 6.3) is conveniently characterized by the following relationship

$$\dot{\epsilon} = \frac{AGDb}{kT}\left(\frac{b}{d}\right)^p\left(\frac{\sigma}{G}\right)^n$$

where A is a dimensionless constant, G is the shear modulus, D is the diffusivity, b is the Burgers vector of the dislocations taking part in the creep process, d is the grain size, p is the inverse grain size exponent, n is the stress exponent, k is Boltzmann's constant, and T is the temperature in kelvin. The diffusion coefficient, D, can be written as

$$D = D_0 \exp\left(-\frac{Q}{kT}\right)$$

where D_0 is a frequency-dependent factor and Q is the activation energy.

6.3   Natural ceramic fibers

Although naturally occurring ceramic or mineral fibers are not as common as natural polymeric fibers, there are two important examples, asbestos and basalt fibers. Asbestos fiber has some health hazards associated with it which have curbed its use. We provide a summary of these two natural ceramic fibers before describing synthetic ceramic fibers in more detail.

Asbestos fiber

The term asbestos represents several naturally occurring minerals which are silicate based and are fibrous in form. This fibrous inorganic material has some unique physical and chemical characteristics (Skinner *et al.*, 1988). They have crystalline structure and are very resistant to heat, acids, alkalis and other chemicals. The most common variety is a hydrated magnesium silicate, called chryostile with the chemical formula of $Mg_3(Si_2O_5)(OH_4)$. Asbestos fibers have relatively low strength but they are not attacked by insects or micro-organisms as is the case with vegetable fibers. Their tensile strength varies from 0.5 to 3 GPa while the Young's modulus can vary from 150 to 180 GPa. Aveston (1969) analyzed the strength of various asbestos fibers by considering them to be a fibrous composite made up of 100% fibers, but assuming that these fibers contain flaws and that these fibers are held together by friction.

The use of asbestos is being curbed because it has been shown to cause lung cancer if inhaled. A simple explanation for this is that the straight molecules in asbestos cleave easily. The asbestos dust consists of small sized fibrous particles that have a large aspect ratio (length/diameter). The consensus of opinion is that these tiny fibers become embedded in the soft tissue of human lungs, causing the tumor. Some respiratory diseases are also caused by inhalation of asbestos fibers. Cooper (1971) summarized the different illnesses associated with occupational exposure to asbestos such as asbestosis, pleural calcification, and cancers of the pulmonary and gastrointestinal tracts and thoracic cavity. Nonoccupational exposure to asbestos can also cause illnesses. Asbestos has been used as an insulation material in homes, schools, brake linings of cars, hair driers, and in many other products. In most countries, legislation defines the maximum allowable exposure to asbestos fibers in the workplace.

Since extremely small diameter fibers, diameter between 1 and 3.5 μm, can be respirated by human beings, it would appear that one way to reduce this risk is to increase the fiber diameter. For example, an early process of making refractory fibers for furnace linings involved blowing high pressure air or stream to attenuate the molten drops of alumina-silica. The fiber diameter range obtained in this operation was from submicrometer to 10 μm, with an average of about 3 μm (Deren, 1995). Modern techniques involve the spinning of fibers which has shifted the diameter of fibers to larger sizes (length and diameter). The average fiber diameter produced by a centrifugal technique that accelerates the fibers away from the molten stream is about 4 μm. It should be remembered, however, that both blowing and spinning techniques produce some submicrometer fibers that are in the respirable range of human beings. There have been attempts to convert chryostile asbestos into a harmless variety. For example, Mackenzie and Meinhold (1994) developed a process of bonding chryostile asbestos by impregnation with sodium silicate followed by a thermal treatment to convert the same chryostile asbestos into crystalline forsterite. The conversion of chryostile to forsterite reduced the respirable fiber concentration and destroyed the cell toxicity of raw chryostile. These glass bonded materials showed good resistance to mechanical abrasion.

Basalt fiber

Basalt is an igneous rock found in North America, Eastern Europe, Russia, etc. An approximate chemical composition of basalt (in wt%) is as follows: SiO_2 50, Al_2O_3 15, $(FeO+Fe_2O_3)$ 13, TiO_2 3, MnO 0.2, CaO 9, MgO 5, K_2O 1, Na_2O 3, and $P_2O_5 < 1$. The exact composition depends on the native basalt rock from which the fibers are obtained. Basalt fibers can be obtained by a process similar to that of glass fiber manufacture (see Chapter 7). The basalt rock is melted (1200–1400°C) and fed to electrically heated platinum–rhodium bushings with 200 or more holes. The fiber diameter depends on the melt temperature and the speed of pulling, commonly it is between 10 and 15 μm. Continuous fibers have been drawn through a platinum-rhodium bushing, electrically heated for 2 h at 1325°C. Similar to glass fiber, silanes or other sizes are applied to these fibers before winding on a drum. Basalt fibers made by drawing have strength and modulus high enough for reinforcement of polymeric materials (Subramanian *et al.*, 1977; Jang and Subramanian, 1993).

6.4   Synthetic ceramic fibers

The last quarter of the twentieth century saw tremendous advances in the processing of continuous, fine diameter ceramic fibers. Figure 6.4 provides a summary of some of the important synthetic ceramic fibers that are available commercially. We have included in Fig. 6.4 two elemental fibers, carbon and boron, while we have excluded the amorphous, silica-based glasses. Two main categories of synthetic ceramic fibers are: oxide and nonoxides. A prime example of oxide fibers is alumina while that of nonoxide fibers is silicon carbide. An important subclass of oxide fibers are silica-based glass fibers and we devote a separate chapter to them because of their commercial importance (see chapter 7). There are also some borderline ceramic fibers such as the elemental boron and carbon fibers. Boron fiber is described in this chapter while carbon fiber is described separately, because of its commercial importance, in Chapter 8.

Processing of ceramics into fibrous form requires very high melt temperatures and the melts generally do not have the rheological characteristics suitable for the melt drawing process. Thus, to make these fibers, one commonly has to resort to techniques without the melting step, but invariably with a sintering or firing step. A low firing or sintering temperature will result in a small grain size but is very likely to lead to an unacceptable level of residual porosity. At higher pro-

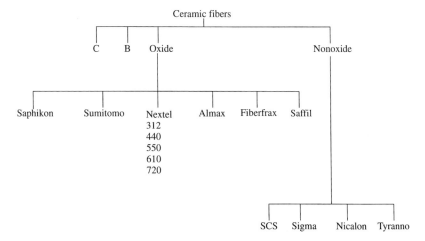

Figure 6.4   Some of the important synthetic ceramic fibers that are available commercially.

cessing temperatures, one can eliminate porosity but excessive grain growth will result because of the high temperatures involved. One can avoid this dilemma by introducing a second phase that restricts grain boundary mobility while the porosity is removed at high temperatures. In principle, it is possible to choose the type and amount of second phase that inhibits grain growth at the service temperature.

6.4.1   Oxide fibers

Crystalline oxide fibers represent an important class of ceramic fibers mainly because of their superior oxidation resistance, being oxides. We describe the processing, structure, and properties of oxide fibers, mainly alumina and some alumina+silica-type fibers.

Alumina-type oxide fibers

Alumina can have γ, δ, η, and α allotropic forms. α-alumina is the thermodynamically stable form. In practice, it is very difficult to obtain the precise time and temperature conditions to proceed from γ to α. Among other problems, internal porosity trapped by grain boundary migration at high processing temperatures is difficult to eliminate completely. The reader should note that for high creep resistance, one requires a large grain size because grain boundary sliding can lead to a large creep strain in a fine-grained material. It is possible to use oxides of silicon, phosphorus, boron or zirconium as inhibitors of grain boundary migration in alumina but generally they lower the working temperature.

Many companies have the capability to produce alumina fibers, polycrystalline or single crystal and in different diameters. Du Pont Co. produced, at one time, a continuous filament, polycrystalline α-alumina yarn, called FP, and later an α-alumina+15–20% ZrO_2 yarn, called PRD-166. 3M Co. makes a range of alumina, alumina+silica, alumina+silica+boria fibers, all under the trade name of Nextel. Sumitomo Chemical Co. produces a fiber that can have a range of composition: 70–100% Al_2O_3 and 30–0% SiO_2. Mitsui Mining Co. produces a polycrystalline, α-alumina fiber called Almax. Continuous, single crystal aluminum oxide or sapphire fibers can be produced by drawing from molten alumina. A commercial fiber produced by this method, called Saphikon, has a hexagonal structure with its c-axis parallel to the fiber axis. The diameter is large, between 75 and 250 μm. ICI produces a staple fiber, called Saffil, that is 96% δ-alumina and 4% silica. Unifrax (formerly Carborundum) produces a series of staple fibers with a composition that is a mixture of alumina and silica in different proportions. Most of such oxide staple fibers are used for thermal insulation purposes. We describe below the salient features of some of the oxide fibers.

α-alumina fiber

Researchers at du Pont (Dhingra, 1980) made a continuous α-alumina fiber by spinning a viscous solution. Du Pont Co. does not make these fibers any more. Nevertheless, we give a brief description of this process of making alumina fiber because it represents an important technique. The basic fabrication procedure involves three steps:

(i) An aqueous slurry mix of selected alumina particles and some additives to render it spinnable is made. For example, aluminum oxychloride [$Al_2(OH)_5Cl$] can be used to produce a solution containing a high oxide-equivalent in the precursor for alumina fibers. The viscosity of this slurry is controlled by controlling the amount of water present.

(ii) Fibers are dry spun from this spinnable slurry.

(iii) The dry-spun yarn is subjected to a two-step firing. This two step firing is done in order to have controlled shrinkage during the transformation to a crystalline ceramic fiber. The first step of low firing controls the shrinkage, while the second step, called flame firing, improves the density of α-Al_2O_3 (98% of theoretical density). A thin silica coating applied to the surface of this fiber can serve to heal the surface flaws, giving about 50% higher tensile strength than the uncoated fibers.

Figure 6.5a, a TEM micrograph, shows the grains in this continuous α-alumina fiber (trade name FP fiber). A modification of the FP alumina, called PRD-166 fiber, was also made by du Pont (Romine, 1987, Nourbakhsh *et al.*, 1989). PRD-166, an α-alumina fiber, about 20 μm in diameter, had 15–20 wt% yttria-stabilized zirconia particles. Figure 6.5b shows the microstructure of the PRD-166 fiber as seen in a scanning electron microscope. The fibers had an average grain size of about 0.5 μm, with the zirconia particles (about 0.1 μm in diameter) located mostly at grain boundary triple points. Their function is to inhibit grain growth in alumina. The small, white particles in Fig. 6.5b are zirconia.

Although oxide fibers have, in general, superior high temperature stability in air *vis-à-vis* nonoxide fibers, these fibers lose their strength because of grain growth. A stable grain structure would be desirable for superior creep resistance at high temperatures. The idea of incorporation of second phase particles is to take care of this problem. In this regard, a new fiber, Nextel 720, having a two-phase microstructure (mullite + α-alumina) has very good creep resistance (see below).

3M alumina fibers

An important polycrystalline α-alumina fiber has been developed by 3M Co. This α-alumina fiber, trade name *Nextel 610*, is made via a sol–gel route. The sol–gel process of making fibers involves the following steps common to all sol–gel processing:

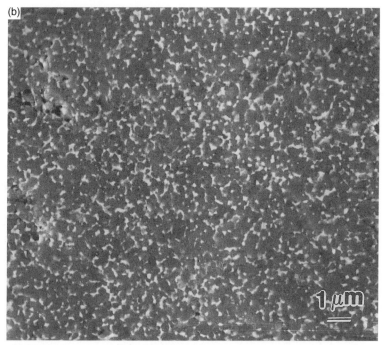

Figure 6.5 (a) Microstructure of an alumina fiber. TEM. (b) Microstructure of an (alumina+zirconia) fiber. SEM. The white particles are zirconia.

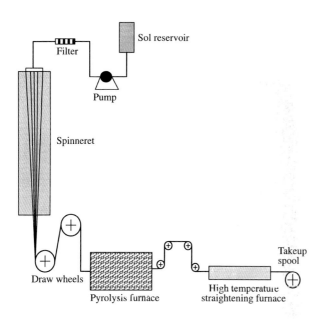

Figure 6.6 Schematic of the 3M process of making alumina fiber.

(a) formulate sol;

(b) concentrate to form a viscous gel;

(c) spin the precursor fiber;

(d) calcine to obtain the oxide fiber.

Specifically, in the case of the 3M process of making Al_2O_3 fiber, the following steps are involved:

- Use an organic basic aluminum salt solution as the starting material.
- Drive out the organics by decomposing and volatilizing without causing cracking, blistering, or other defects.
- Final firing at 1400°C under carefully controlled conditions.
- A low temperature straightening treatment.

Figure 6.6 shows the 3M process schematically.

Wilson (1990) has provided some details of the microstructural evolution in this fiber. A fine-grained α-Al_2O_3 fiber is obtained by seeding the high temperature α-alumina with a very fine hydrous colloidal iron oxide. The fine iron oxide improves the nucleation rate of α-Al_2O_3, with the result that a high density, ultra-fine, homogeneous α-Al_2O_3 fiber is obtained. The rationale for seeding with iron oxide as follows. Basic salts of aluminum decompose into transition aluminum oxide spinels such as η-Al_2O_3 above 400°C. These transition cubic spinels convert to hexagonal α-Al_2O_3 on heating to between 1000 and 1200°C. The problem is that the nucleation rate of pure α-Al_2O_3 is too low and results in large grains. Also, during the transformation to α phase, large shrinkage results in

Figure 6.7   Nextel 610 α-alumina fiber. S E M.

large porosity (Sowman, 1988, Birchall *et al.*, 1985, Saitow *et al.*, 1992). Seeding of alumina with fine particles would appear to be a solution. α-Fe_2O_3 is iso-structural with α-Al_2O_3, with a 6.5% lattice mismatch (Wilson, 1990). 3M uses a hydrous colloidal iron oxide sol as nucleating agent. According to Wilson, without the seeding of iron oxide, the η-alumina to α-alumina transformation occurs at about 1100°C. With 1% Fe_2O_3, the transformation temperature was decreased to 1010°C, while with 4% Fe_2O_3, the transformation temperature came down to 977°C. Concomitantly, the grain size was refined. The Nextel 610 fiber has 0.4–0.7% Fe_2O_3 and about 0.5 wt% SiO_2. SiO_2 is added to reduce the final grain size, although SiO_2 inhibits transformation to the a phase. The SiO_2 addi-tion also reduces grain growth during soaking at 1400°C. Figure 6.7 shows a scanning electron micrograph of the surface of a Nextel 610 (99.5% α-alumina) fiber fired to 1400°C. Note the smooth fiber surface.

Other aluminosilicate fibers

Many other alumina or alumina–silica-type fibers are available. Most of these are made by the sol–gel process. Sumitomo Chemical company produces a fiber that is a mixture of alumina and silica. The flow diagram of this process is shown in Fig. 6.8. Starting from an organoaluminum (polyaluminoxanes or a mixture of polyaluminoxanes and one or more kinds of Si containing compounds), a pre-

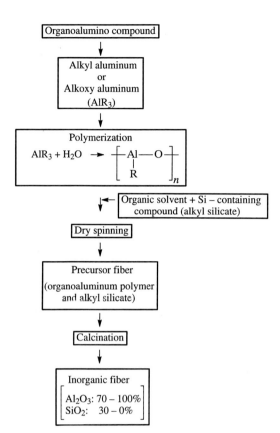

Figure 6.8 Schematic of the Sumitomo process for making an (alumina + silica) fiber.

cursor fiber is obtained by dry spinning. This precursor fiber is calcined to produce the final fiber. The structure and properties of the Sumitomo alumina–silica fiber have been studied by Bunsell *et al.* (1988) and Lesniewski *et al.* (1990). Compositionally, this fiber consists of crystalline gamma alumina + amorphous silica. On heating to 1127°C, mullite formed slowly while at 1400°C, mullite formation occurred rapidly (Lesniewski *et al.*, 1990). Their observations showed that the fiber was stable up to 800°C, and the properties showed a distinct decline in the temperature range 800–1127°C, which was attributed to softening of silica.

The Nextel series of fibers produced by the 3M Company consists of a variety of aluminosilicate fibers. These contain mainly $Al_2O_3 + SiO_2$ and some B_2O_3. The compositions and properties of Nextel 312, Nextel 440, Nextel 550, Nextel 610, and Nextel 720 fibers are given in Table 6.3. Nextel 610 is, of course, the α-alumina fiber described above. The sol–gel manufacturing process used by 3M Co. has metal alkoxides as the starting materials. Metal alkoxides are $M(OR)_n$-type compounds where M is the metal and n is the metal valence, and R is an organic compound. Selection of an appropriate organic group is very important. It should provide sufficient stability and volatility to the alkoxide so M–OR bonds are broken and MO–R are obtained to give the desired oxide ceramics. Hydrolysis of metal alkoxides results in sols which are spun and gelled. The gelled fiber is then densified at relatively low temperatures. The high surface free energy available in the pores of the gelled fiber allows for relatively low temperature densification. The process provides close control over solution composition and rheology of fiber. The disadvantage is that rather large dimensional changes must be accommodated and fiber integrity conserved.

Sowman (1988) has provided details of the process used by 3M Co. for making the Nextel oxide fibers. Aluminum acetate [Al(OH)$_2$(OOCCH$_3$). 1/3H$_3$BO$_3$], e.g. Niaproof, from Niacet Corp., is the starting material. Aluminum acetate having an Al$_2$O$_3$/B$_2$O$_3$ ratio of 3 to 1 becomes spinnable after water removal from an aqueous solution. In the fabrication of 3M continuous fibers, a 37.5% solution of basic aluminum acetate in water is concentrated in a rotating flask partially immersed in a water bath at 32–36°C. After concentration to an equivalent Al$_2$O$_3$ content of 28.5%, a viscous solution with viscosity, η, between 100 and 150 Pa s is obtained. This is extruded through a spinneret having 130 holes of 100 μm diameter under a pressure of 800–1000 kPa. Shiny, colorless fibers are obtained on firing to 1000°C. The microstructure shows cube-shaped and lath-shaped crystals. The boria addition lowers the temperature required for mullite formation and retards the transformation of alumina to α-Al$_2$O$_3$. One needs boria in an amount equivalent to or greater than a 9 Al$_2$O$_3$: 2 B$_2$O$_3$ ratio in Al$_2$O$_3$–B$_2$O$_3$–SiO$_2$ compositions to prevent the formation of crystalline alumina.

An important fiber in this series is Nextel 720. It is also a sol–gel processed fiber with the composition 85% Al$_2$O$_3$–15% SiO$_2$ (Wilson et al., 1995). The distinctive feature of this fiber is that it has a two-phase crystalline microstructure consisting of globular and elongated grains (<100 nm) of α-Al$_2$O$_3$ and mullite

Table 6.3 *Composition and properties of Nextel series fibers.*

Fiber type	Composition (wt %)	Diameter (μm)	Density (g cm^{-3})	Tensile strength (MPa)	Young's modulus (GPa)
Nextel 312	Al$_2$O$_3$–62, SiO$_2$–24, B$_2$O$_3$–14	10–12	2.7	1700	152
Nextel 440	Al$_2$O$_3$–70, SiO$_2$–28, B$_2$O$_3$–2	10–12	3.05	2000	186
Nextel 550	Al$_2$O$_3$–73, SiO$_2$–27	10–12	3.03	2000	193
Nextel 610	Al$_2$O$_3$–99+, SiO$_2$–0.2–0.3, Fe$_2$O$_3$–0.4–0.7	10–12	3.75	1900	370
Nextel 720	Al$_2$O$_3$–85, SiO$_2$–15	10–12	3.4	2130	260

Figure 6.9 An 8-harness satin weave fabric of Nextel 720 fiber (courtesy of N. Chawla)

1.5 mm

$(3Al_2O_3.2SiO_2)$ and large, globular mosaic crystals $(0.5\,\mu m)$ of mullite and α-alumina. The mosaic structure consists of multiple grains having similar crystal orientations and separated by low angle grain boundaries. The final ceramic fiber is obtained by firing at 1350°C to convert the 85% alumina + 15% SiO_2 (amorphous) composition to approximately 59 vol% mullite and 41 vol% Al_2O_3.

The small diameter of all Nextel fibers allows them to be easily woven. As an example, Fig. 6.9 shows an 8-harness satin weave fabric of Nextel 720 fiber.

Other alumina fibers

A δ-Al_2O_3 fiber produced by the sol–gel process became commercially available in the 1970s. This fiber is short in length (3–5 mm) and has a very fine diameter $(3\,\mu m)$. In textile terminology, such a short fiber is called a *staple* fiber. This staple fiber carries the trade name 'Saffil'. It contains about 4% SiO_2 which serves to stabilize the δ-alumina phase and also inhibits grain growth.

A sol–gel method is used to produce silica-stabilized alumina (Saffil) and calcia-stabilized zirconia fibers. The flow diagram for Saffil fiber is shown in Fig. 6.10. For Saffil alumina fiber, aluminum oxychloride $[Al_2(OH)_5Cl]$ is mixed with a medium molecular weight polymer such as 2 wt% polyvinyl alcohol. The aqueous phase contains an oxide sol and an organic polymer. The sol is extruded as filaments into a coagulating (or precipitating) bath in which the extruded shape gels. The gelled fiber is then dried and calcined to produce the final oxide fiber. This solution is slowly evaporated in a rotary evaporator until a viscosity

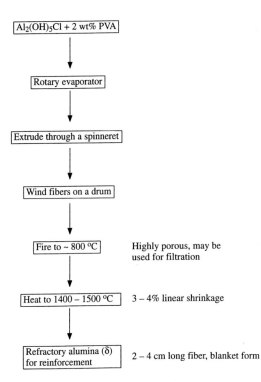

Figure 6.10   Flow diagram of sol–gel processed δ-alumina, staple fibers (Saffil).

of about 80 Pa s (800 poise) is attained. This solution is extruded through a spinneret, and the fibers are wound on a drum and fired to about 800°C. The organic material is burned away and a fine-grained alumina fiber having 5–10% porosity and a diameter of 3–5 μm is obtained. The fibers produced at this stage are suitable for filter purposes because of their high porosity. By heating them to 1400–1500°C which causes a 3–4% linear shrinkage, one obtains a refractory alumina fiber suitable for reinforcement purposes.

Single crystal oxide fiber

A technique called *edge-defined film-fed growth* (EFG) has been used to make continuous, monocrystalline sapphire (Al_2O_3) (LaBelle and Mlavsky, 1971; Labelle, 1971; Pollack, 1972; Hurley and Pollack, 1972). LaBelle and Mlavsky (1971) were the first to grow sapphire (Al_2O_3) single crystal fibers using a modified Czochralski puller and radio frequency heating. In 1971, these authors devised a growth method, called the edge-defined, film-fed growth (EFG) method. Figure 6.11 shows a schematic of the EFG method. Growth rates as high as 200 mm/min have been attained. The die material must be stable at the melting point of alumina. A molybdenum or tungsten die is commonly used. The technique has also been successfully used to make continuous fibers of yttrium aluminum garnet (YAG) and yttria alumina eutectic.

In this technique, a capillary supplies a constant liquid level at the crystal

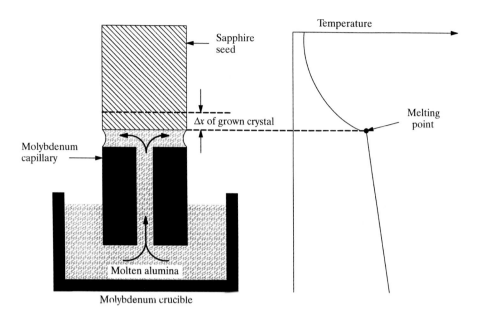

Figure 6.11 Schematic of the edge-defined film-fed growth (EFG) method of making a single crystal alumina fiber.

interface. A sapphire seed crystal is used. Molten alumina wets both, molybdenum and alumina. The crystal grows from a molten film between the growing crystal and the die. The crystal shape is defined by the external shape of the die rather than the internal shape.

A *laser-heated floating zone method* has been devised to make a variety of ceramic fibers. Gasson and Cockayne (1970) used laser heating for crystal growth of Al_2O_3, Y_2O_3, $MgAl_2O_4$ and Na_2O_3. Haggerty (1972) used a four-beam laser-heated float zone method to grow single crystal fibers of Al_2O_3, Y_2O_3, TiC, and TiB_2. The laser-heated float zone technique is shown in Fig. 6.12. A CO_2 laser is focused on the molten zone. A source rod is brought into the focused laser beam. A seed crystal, dipped into the molten zone, is used to control the orientation. Crystal growth starts by moving the source and seed rods simultaneously. Mass conservation dictates that the diameter is reduced as the square root of the feed rate/pull rate ratio. It is easy to see that, in this process, the fiber purity is determined by the purity of the starting material. Continuous alumina fiber and alumina/YAG ($Y_3Al_5O_{12}$) eutectic fibers have been made by laser-heated floating-zone method (Sayir and Farmer, 1995; Sayir *et al.*, 1995). The alumina/YAG eutectic fibers are obtained by directional solidification to align the components along the fiber axis. A slurry is made of a mixture in eutectic proportion (18.5 mol% yttria and 81.5 mol% alumina) in water plus a methylcellulose binder, and glycerine as a plasticizer. The slurry is dried to paste and extruded into a fiber. A rod is made of the extruded fiber and used as feedstock for the

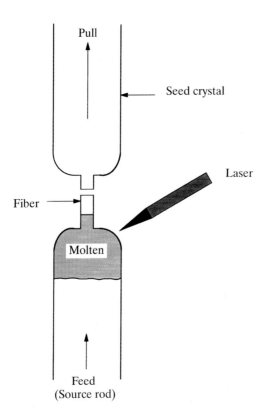

Pull

Seed crystal

Laser

Fiber

Molten

Feed
(Source rod)

Figure 6.12 Schematic of the laser-heated float zone technique of making ceramic fibers.

laser-heated floating-zone method. A CO_2 laser is focused on the fiber as it is drawn, creating a small melt zone that moves along the fiber. As the melt zone moves, the resolidified fiber left behind has alternating lamellae of alumina and YAG. The thinner the lamellae, the stronger is the resultant fiber. The thickness of the lamellae can be controlled by the rate of solidification, the diameter of the fiber, and the thermal gradient in the melt zone.

Yet another novel technique of making oxide fibers is called the *inviscid melt technique*. It is true that, in principle, any material that can be made molten can be drawn into a fibrous shape. Organic polymeric fibers such nylon, aramid, etc. as well as a variety of glasses are routinely converted into a fibrous form by passing a molten material, having an appropriate viscosity, through an orifice. The inviscid (meaning *low viscosity*) melt technique uses this principle and a schematic of the method is shown in Fig. 6.13a. Essentially, the technique involves the extrusion of a low-viscosity molten jet through an orifice into a chemically reactive environment. The low-viscosity jet is unstable with respect to surface tension because of a phenomenon called Rayleigh waves. Here, perhaps, it will be in order to digress a little and say a few words about the so-called surface or Rayleigh waves. Rayleigh waves are surface waves (Savart, 1833, Rayleigh 1879). The reader will recall that the two common modes of wave motion in solids

(a)

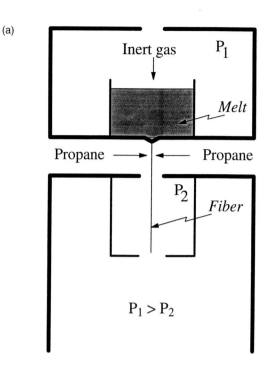

Figure 6.13 (a) Inviscid melt technique of making ceramic fibers. For (b) and (c) see overleaf.

are longitudinal and shear waves. In a longitudinal (or compressional wave), the particle moves parallel to the wave propagation direction while in a shear wave, the particle moves perpendicular to the direction of wave propagation. Surface waves are present in a thin surface layer of a solid. There are two types of surface waves: Rayleigh and Lamb waves. Surface waves of interest to us here are the Rayleigh waves. Rayleigh waves, sometimes referred to as R-waves, are also important in seismology. They are waves on the surface of the earth with a retrograde, elliptical motion at the force surface. Rayleigh waves are confined to a thin surface layer of a solid and the particle trajectories are ellipses in planes normal to the surface and parallel to the direction of wave propagation. These are the surface waves that form on the surface of the low-viscosity jet stream. The study of this phenomenon of instability of a liquid jet can be traced to Savart and later to Rayleigh in the nineteenth century. When a fluid is forced through an orifice under pressure, Rayleigh or surface waves grow exponentially in amplitude and tend to breakup the jet into a series of droplets. Generally, these droplets will be of random size. If, however, we can apply a regular perturbation of the right frequency to the jet, we can impose a standing wave on the fluid cylinder coming out of the orifice (Bogy, 1979). This will result in breakup of the jet in a controlled and periodic manner, and we will have equally spaced droplets whose size is very uniform and reproducible. The droplet or particle size depends on the original jet diameter and the wavelength of the imposed mechanical perturbance. In the case of fiber formation, our objective is to stabilize the molten jet against breakup by

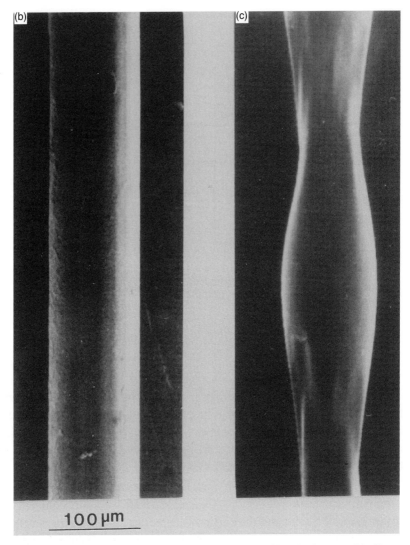

Figure 6.13 (*cont.*)   (b) A stabilized, round inviscid melt spun fiber (c) An unstable fiber with Rayleigh waves frozen on its surface, a condition just before breakup (courtesy of F. Wallenberger).

the Rayleigh waves. In the case of glasses and organic polymers, the melts have high viscosity (more than 10^4 Pa s or 10^5 poise), and the high viscosity delays the Rayleigh breakup until the molten jet freezes. In the case of a low-viscosity melt, one can avoid breakup of the molten jet by chemically stabilizing it. For example, in a process used at Du Pont Co. to make fibers of alumina + calcia, the low-viscosity (1 Pa s or 10 poise) jet was chemically stabilized with propane before the Rayleigh waves could break the stream into droplets. Figure 6.13b shows a stabilized, round inviscid melt spun fiber while Fig. 6.13c shows a fiber with Rayleigh waves frozen on its surface, a condition just before breakup (Wallenburger *et al.*, 1992). The inviscid jet must be stabilized in about 10^{-3} seconds or it will break up into droplets. Yet another point in regard to this process is the fact that small diameter fibers are difficult to make. For example, the smallest diameter of an alumina–calcia fiber produced by this technique is 105 μm. These fibers, according to the authors, have tensile strengths close to 1 GPa.

Properties of oxide fibers

The composition and some of the physical and mechanical characteristics (at room temperature) of oxide fibers are summarized in Tables 6.3 and 6.4. Oxide fibers, being oxides, are inherently stable in air at high temperatures. They are, however, prone to creep deformation at high temperatures (Gooch and Grover, 1973; Jakus and Tulluri, 1989; Johnson *et al.*, 1987; Pysher and Tressler, 1992). The creep rate of fine-grained alumina-based fibers is controlled by grain boundary diffusion, characterized by a linear dependence of creep rate on stress. In single crystal alumina fibers, creep occurs by dislocation slip or pyramidal slip (see Fig. 6.2), which results in a high stress exponent. A comparison of various polycrystalline alumina and aluminosilicate fibers in the Nextel series is shown in Fig. 6.14. Note the superior creep resistance of the Nextel 720 fiber, with the stress exponent n being equal to 3. Nextel 610 fiber also showed $n=3$. The steady state creep rate, however, was lower for the 720 fiber. The value of $n=3$ would indicate that the operative creep mechanism is grain boundary or interface controlled. Wilson *et al.* (1995) attributed this superior creep resistance to the superior creep resistance of mullite compared to alumina and to a two-phase, mosaic microstructure consisting of elongated and globular grains, resulting in a bridging effect due to the elongated grains similar to that observed in silicon carbide whisker reinforced alumina (Chawla, 1993). When fine-grained fibers are held at high temperatures, their grain size increases and, consequently, their room temperature strength decreases. For example, Xu *et al.* (1993) observed loss of strength in Nextel 610 alumina fibers upon isothermal exposure due mainly to grain growth. Similar grain size growth was observed in Nextel 720 fibers (Schneider *et al.*, 1996). In the case of Nextel 720, however, we have alumina and mullite grains. Figure 6.15 shows the evolution of microstructure of Nextel 720 fiber as a function of isothermal exposure at different temperatures. Quantitative information on the increase in grain size of mullite and alumina with temperature is shown in Fig. 6.16a while the decrease in the residual strength of Nextel 720 with increasing annealing temperature (i.e. increasing grain size) is shown in Fig. 6.16b. F5 and F100 in Fig. 6.16b refer to the 5 and 100 mm gage lengths used in tensile testing of these fibers.

Morscher and Sayir (1995) studied the effect of temperature on the bend radius that a c-axis-oriented sapphire fiber can withstand for fibers of various diameters. They did this by performing bend stress rupture tests on these fibers

Table 6.4 *Composition and properties of some non-Nextel oxide fibers.*

Trade name	Manufacturer	Composition (wt %)	Diameter (μm)	Density, ρ (g cm^{-3})	Tensile strength, (GPa)	Tensile modulus, E (GPa)	Coefficient of thermal expansion, α (10^{-6} K^{-1})	Strain to failure ϵ (%)	Process
Sumica	Sumitomo Chemical	Al$_2$O$_3$–85, SiO$_2$–15	9	3.2	2600	250	—	—	Sol-gel
Saphikon	Saphikon	Al$_2$O$_3$–100%	75–225	3.97	2.1–3.4	386–435	7.9–8.8 11.5 $\parallel c$ axis		Melt grown
Almax	Mitsui Mining	Al$_2$O$_3$>99%	15	3.6	1.8	320	—	0.5	Slurry spun
Saffil	ICI	Al$_2$O$_3$–96, SiO$_2$–4	3	2.3	1000	100	—	—	Sol-gel
Altex	Sumitomo Chemical	Al$_2$O$_3$–95% SiO$_2$–5%	10 15	3.3	1.8	210	8.8	0.8	Sol-gel

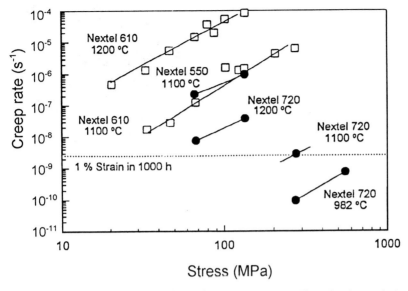

Figure 6.14 A comparison of various polycrystalline alumina and aluminosilicate fibers in the Nextel series. Note the superior creep resistance of the Nextel 720 fiber. 3M data.

for times of 1–100 h and temperatures of 300–1700°C. Fibers would show one of the following three behaviors: survive the bend test undeformed, fracture or deform. The radius above which no fibers fractured or deformed for a given time–temperature treatment was the bend survival radius. The ability of fibers to withstand curvature decreased substantially with time and increasing temperature, and fibers of smaller diameter (40–83 μm) withstood smaller bend radii than would be expected from just a difference in fiber diameter when compared with the bend results of the fibers of large diameter (144 μm). The authors attributed this to different flaw populations, causing high temperature bend failure for the tested sapphire fibers of different diameters.

6.4.2 Nonoxide ceramic fibers

Ceramic fibers of the nonoxide variety such as silicon carbide, silicon oxycarbide such as Nicalon, silicon nitride, boron carbide, etc. have become very important because of their attractive combination of high stiffness, high strength and low density. We give brief description of some important nonoxide fibers.

Silicon carbide fibers

Silicon carbide fiber must be regarded as a major development in the field of ceramic fibers during the last quarter of the twentieth century. In particular, a process developed by the late Professor Yajima in Japan, involving controlled

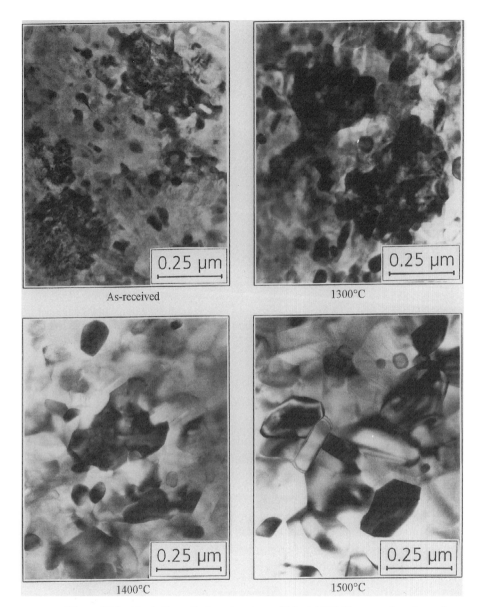

Figure 6.15   Evolution of microstructure of Nextel 720 fiber as a function of isothermal exposure at different temperatures (courtesy of H. Schneider and M. Schmücker)

pyrolysis of a polycarbosilane precursor to yield a flexible fiber, must be considered to be the harbinger of the making of ceramic fibers from polymeric precursors. Pyrolysis of a suitable organometallic to produce an inorganic ceramic material must be regarded as one of the greatest success stories of the twentieth century. In particular, as we shall see later, polysilane and polycarbosilane are useful precursors for silicon carbide while polysilazanes are suitable for silicon

(a)

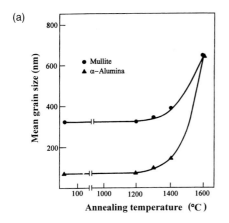

(b)

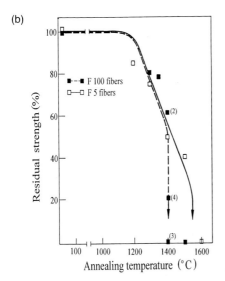

Figure 6.16 (a) Increase in grain size of mullite and alumina in Nextel 720 fiber as a function of the annealing temperature. (b) Drop in residual strength of Nextel 720 as function of the increasing annealing temperature (after Schneider *et al.*, 1996). F5 and F100 in Fig. 6.16b refer to the 5 and 100 mm gage lengths used in tensile testing of these fibers.

nitride and silicon carbonitride. In this section we describe the processing, microstructure, and properties of these and other silicon carbide fibers in some detail.

We can easily classify the fabrication methods for SiC as conventional and nonconventional. The former category would include chemical vapor deposition while the latter would include controlled pyrolsis of polymeric precursors. There is yet another important type of SiC available for reinforcement purposes, SiC whiskers. We give a brief description of these.

CVD silicon carbide fibers

Silicon carbide can be made by chemical vapor deposition on a substrate heated to around 1300°C (DeBolt *et al.*, 1974). The substrate can be tungsten or carbon. The reactive gaseous mixture contains hydrogen and alkyl silanes. Typically, a gaseous mixture consisting of 70% hydrogen and 30% silanes is introduced at the

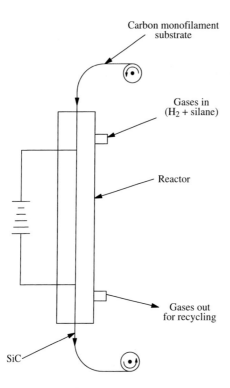

Carbon monofilament
substrate

Gases in
(H₂ + silane)

Reactor

Gases out
for recycling

SiC

Figure 6.17 Chemical vapor deposition (CVD) process of making silicon carbide fiber.

reactor top, Fig. 6.17, where the tungsten substrate (~13 μm diameter) also enters the reactor. Mercury seals are used at both ends as contact electrodes for the filament. The substrate is heated by combined direct current (250 mA) and very high frequency (VHF~60 MHz) to obtain an optimum temperature profile. To obtain a 100 μm SiC monofilament, it generally takes about 20 s in the reactor. The filament is wound on a spool at the bottom of the reactor. The exhaust gases (95% of the original mixture + HCl) are passed around a condenser to recover the unused silanes. Efficient reclamation of the unused silanes is very important for a cost effective production process. This CVD process of making SiC fiber is very similar to that of B fiber. The nodules on the surface of SiC are smaller than those seen on B fibers. Such CVD processes result in composite monofilaments which have built in residual stresses. The process is, of course, very expensive. Methyltrichlorosilane is an ideal raw material as it contains one silicon and one carbon atom, i.e. one would expect a stoichiometric SiC to be deposited. The chemical reaction is:

$$CH_3SiCl_3 \; (g) \xrightarrow{H_2} SiC(s) + 3 \; HCl(g)$$

An optimum amount of hydrogen is required. If the hydrogen is less than sufficient, chlorosilanes will not be reduced to Si and free carbon will be present

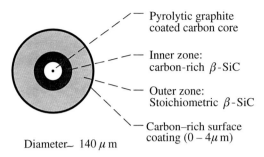

Pyrolytic graphite
coated carbon core

Inner zone:
carbon-rich β-SiC

Outer zone:
Stoichiometric β-SiC

Carbon–rich surface
coating (0 – 4μ m)

Diameter $\sim$ 140 μ m

Figure 6.18 Structure of the SCS-6 silicon carbide fiber and its characteristic surface compositional gradient.

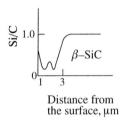

Distance from
the surface, μm

in the mixture. If too much hydrogen is present, excess Si will form in the end product. Generally, solid (free) carbon and solid or liquid silicon are mixed with SiC. The final monofilament (100–150 μm) consists of a sheath of mainly β-SiC with some α-SiC on the tungsten core. The {111} planes in SiC deposit are parallel to the fiber axis. Figure 6.18 shows schematically the SCS silicon carbide fiber and its characteristic surface compositional gradient.

Another CVD type silicon carbide fiber, available commercially, is called *Sigma fiber*. Sigma fiber filament is a continuous silicon carbide monofilament obtained by CVD on a tungsten substrate. Figure 6.19 gives the flow diagram for the fabrication of this process.

Nonoxide fibers via polymers

As pointed out above, the SiC fiber obtained via CVD is very thick, and it is consequently not very flexible. Work on alternative routes for obtaining fine, continuous, and flexible fiber had been in progress for some time when in the mid-1970s the late Professor Yajima and his colleagues (Yajima *et al.*, 1976, Yajima, 1980) in Japan developed a process of making such a fiber by controlled pyrolysis of a polymeric precursor. This method of using silicon-based polymers to produce a family of ceramic fibers having good mechanical properties, good thermal stability, and oxidation resistance has enormous potential. The various steps involved in this polymer route are shown in Fig. 6.20. In any process involving ceramic fiber fabrication via a polymer route, the following are the critical steps (Wax, 1985):

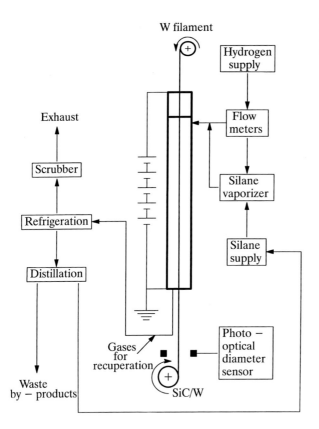

Figure 6.19   Flow diagram for fabrication of the Sigma fiber.

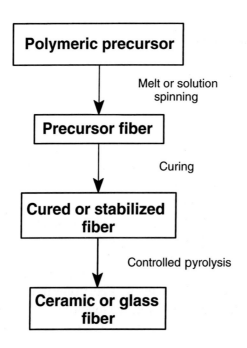

Figure 6.20   Steps involved in the fabrication of ceramic fibers via the polymer route.

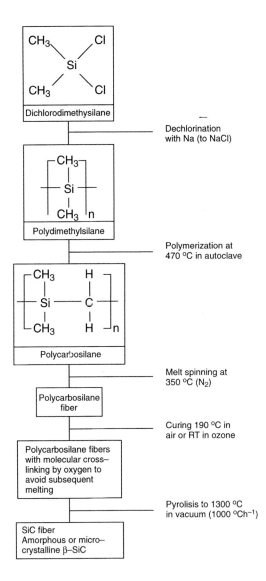

Figure 6.21 Yajima process of making Nicalon (Si–C–O) fiber.

(i) polymer characterization (yield, molecular weight, purity, etc.);

(ii) melt spin polymer into a precursor fiber;

(iii) cure the precursor fiber to cross-link the molecular chains, making it infusible during the subsequent pyrolysis. This step may not be required under certain conditions, see below;

(iv) controlled pyrolysis.

Specifically, the Yajima process of making SiC involves the following steps, shown schematically in Fig. 6.21. Polycarbosilane, a high molecular weight polymer containing Si and C, is synthesized. The starting material is commercially available dimethylchlorosilane. Solid polydimethylsilane is obtained by dechlorination of dimethylchlorosilane by reacting it with sodium. Polycarbosilane is obtained by thermal decomposition and polymerization of polydimethylsilane. This is carried out under high pressure in an autoclave at 470°C in an argon atmosphere for 8–14 h. A vacuum distillation treatment at up to 280°C follows. The average molecular weight of the resulting polymer is about 1500. This is melt spun from a 500-hole nozzle at about 350°C under N_2 gas to obtain the so called pre-ceramic continuous, precursor fiber. The precursor fiber is quite weak (tensile strength ~10 MPa). This is converted to inorganic SiC by curing in air, heating to about 1000°C in N_2 gas, followed by heating to 1300°C in N_2 under tension. This basically is the Nippon Carbon Co. process for the manufacture of Nicalon fibers (Simon and Bunsell, 1984; Ishikawa, 1994). Clearly, small variations exist between this and the various laboratory processes. During pyrolysis, the first stage of conversion occurs at around 550°C when cross-linking of polymer chains occurs. Above this temperature, the side chains containing hydrogen and methyl groups decompose. Fiber density and mechanical properties improve sharply. The conversion to SiC occurs above about 850°C.

The multifilament fiber (10–20 μm diameter) as commercially produced consists of a mixture of β-SiC, free carbon and SiO_2. The properties of this fiber are summarized in Table 6.5. The properties of Nicalon start to degrade at temperatures above about 600°C because of the thermodynamic instability of composition and microstructure. A ceramic grade of Nicalon, called Hi Nicalon, having low oxygen content is also available. Yet another version of a multifilament silicon carbide fiber is Tyranno, produced by Ube Industries, Japan. This is made by pyrolysis of poly (titano carbosilanes) and contains between 1.5 and 4 wt% titanium.

Fine-diameter, polymer-derived silicon carbide fibers generally have high

Table 6.5 *Properties of two Nicalon-type fibers.[a]*

Characteristic	Nicalon	Hi-Nicalon
Composition (wt %)	58 Si, 31 C, 11 O	63.7 Si, 35.8 C, 0.5 O
Density (g cm^{-3})	2.55	2.32
Tensile strength (GPa)	2.96	2.80
Young's modulus (GPa)	192	269
Coefficient of thermal expansion ($10^{-6}K^{-1}$)	4	—

Note:
[a] Manufacturer's data.

oxygen content. This results from the curing of the precursor fibers in an oxidizing atmosphere to introduce cross-linking. Cross-linking is required to make the precursor fiber infusible during the subsequent pyrolysis step. One way around this is to use electron beam curing. Other techniques involve dry spinning of high molecular weight carbosilane polymers (Sacks *et al.*, 1995). In this case, the as spun fiber does not require a curing step because of the high molecular weight polycarbosilane polymer used, i.e. it does not melt during pyrolysis without requiring curing. Sacks *et al.* (1995) used dopant additions to produce low oxygen, near-stoichiometric, small diameter (10–15 μm) SiC fibers. They reported an average tensile strength for these fibers of about 2.8 GPa.

There is great deal of interest in polymer-derived SiC fibers with improved thermomechanical stability. An important requirement in making fibers via pre-ceramic polymer processing is to have sufficient spinnability to make the precursor fiber. This route provides small diameter fibers that are flexible. As described above, two such fine diameter fibers that are commercially available are Nicalon and Tyranno. Both are produced by melt spinning. Both of them also show low thermal stability because of their C-rich stoichiometry and high oxygen content. Such fibers suffer degradation by carbothermal reduction type reactions. Volatiles such as CO and SiO are given out and porosity type flaws grow and rapid grain growth occurs. By reducing the oxygen content, one can improve the thermal stability. Hi-Nicalon and Hi-Nicalon type S are improved versions of Nicalon produced by the Nippon Carbon Co. In order to reduce the oxygen content, electron beam radiation curing is employed in making these fibers.

Professor Sacks and his colleagues at the University of Florida have produced a C-rich and near stoichiometric SiC, low oxygen content by dry spinning (Sacks *et al.*, 1995; Toreki *et al.*, 1994). They start with polydimethylsilane (PDMS) which has a Si–Si backbone. This is subjected to pressure pyrolysis to obtain polycarbosilane (PCS) which has a Si–C backbone. The key point in their process is to have a molecular weight of PCS between MW 5000 and 20 000. This is a high MW compared to that used in other processes. The spinning dope is obtained by adding suitable spinning aids and a solvent. This dope is dry spun to produce what are called the green fibers, which are heated under controlled conditions to produce SiC fiber.

Let us reconsider the basics of making silicon carbide fiber from polycarbosilane (PCS), the starting polymer for making silicon carbide fiber. It is a very suitable precursor for making SiC inasmuch as there is one carbon atom for each silicon atom and thus stoichiometric silicon carbide is easy to obtain in the final ceramic fiber. It is, however, not the ideal precursor because it is not quite a linear chain polymer, but has some side groups. The key processing variable here is the molecular weight, MW of PCS. For MW <5000, it is highly soluble, melts easily, and the ceramic yield is low (60–65%). This melt spinning technique is used by Nippon Carbon Co. Thus, the strength of Nicalon green fiber is very low. This makes it difficult to handle. When PCS has a molecular weight >25 000, it does not melt, has low solubility, but the ceramic yield is high. At such high values of MW, it is not very useful for making the precursor fiber. However, for a molecular weight in the 5000–20 000 range, the PCS *does not* melt, is highly soluble, and results in a high ceramic yield (~80%). Thus, in this MW range, an appropriate spinning dope can be used to prepare the green fiber which will not require the curing step. Sacks *et al.* (1995) took this route. A proper polymer/solvent ratio in the spinning dope is required. If there is excess solvent, then not all the solvent will be removed and fibers will stick together. To control the amount of solvent, a flowing gas in a heated column is used. If there is less than adequate solvent, then it will be difficult to extrude the dope, and a rough fiber surface results. The dope viscosity, η should be between 10 and 100 Pa s, nearly Newtonian but with a slight shear thinning characteristic but no anisotropy. Certain processing and spinning aids are, invariably, used. For example, boron addition enhances densification and controls grain growth. It is very important to keep the number of fiber breaks/minute/spinneret hole to a minimum. The green fiber obtained in the University of Florida process has a strength between 18 and 21 MPa for a MW of 7000. If the molecular weight is increased to 16 500, the strength of the green fiber increases to 24 MPa. The final diameter of the fiber is between 10 and 15 μm. The structure of this fiber depends on the final heat treatment temperature. If the heat treatment is between 750 and 1000°C, an amorphous structure (very weak crystallinity) results and the fracture surface shows a typical glassy appearance. When heat treated at 1600°C, more crystallites form but the crystallite size is less than 10 μm. After heat treatment at 1800°C, β-SiC (cubic) peaks appear. According to Sacks *et al.*, elemental analysis shows their fibers to be stoichiometric SiC or slightly C-rich. They report a tensile modulus of 430 GPa, only slightly less than that of 100% β-SiC. Their density measurements showed a $\rho_{avg} = 3.15(\pm 0.04)$ g cm^{-3} while the theoretical density of SiC is $\rho_{theo} = 3.21$ g cm^{-3}. The fact that $\rho_{avg} < \rho_{theo}$ is attributed to the presence of some porosity (<μm) in the fiber, plus some free C (in graphitic form) around the grains.

Laine and coworkers (1993, 1995) and Zhang *et al.* (1994, 1995) at University of Michigan have used a polymethylsilane (PMS) ($-[CH_3SiH]_x-$) as the precursor polymer for making a fine diameter, silicon carbide fiber. Spinning aids were used to stabilize the polymer solution and the precursor fiber was extruded from a 140 μm orifice extruder into an argon atmosphere. The precursor fiber was pyrolyzed at 1800°C in Ar. Boron, added as a sintering aid, helped to obtain a dense product. Figure 6.22 shows a schematic of the University of Michigan process of making SiC fiber.

Another silicon carbide multifilament fiber, made via a polymeric precursor by Dow Corning Corp., USA, is called *Sylramic*. According to the manufacturer, this textile grade silicon carbide fiber has a nanocrystalline, stoichiometric

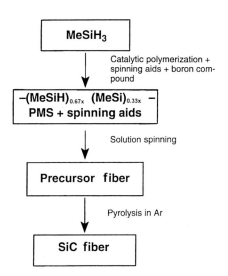

Figure 6.22 Schematic of the University of Michigan process of making SiC fiber (after Laine *et al.*, 1995)

silicon carbide (crystallite size of 0.5 μm). Its density is 3.0 g cm^{-3}, and it has a tensile strength and modulus of 3.15 GPa and 405 GPa, respectively.

Structure and properties of nonoxide fibers

Textron Specialty Materials Co. has developed a series of surface-modified silicon carbide fibers, called SCS fibers. These special fibers have a complex through-the-thickness gradient structure. SCS-6, for example, is a thick fiber (diameter = 142 μm) and is produced by chemical vapor deposition of silicon and carbon-containing compounds onto a pyrolytic graphite coated carbon core, see Fig. 6.18. The pyrolytic graphite coating is applied to a carbon monofilament to give a substrate of 37 μm thickness. This is then coated with SiC by CVD to give a final monofilament of 142 μm diameter. The surface modification of the SCS fibers consists of the following: the bulk of the 1 μm thick surface coating consists of C doped Si. Zone I at and near the surface is a carbon rich zone. In zone II, Si content decreases. This is followed by a zone III in which the Si content increases back to the stoichimetric SiC composition. Thus, the SCS silicon carbide fiber has a surface graded outward to be carbon rich and back to stoichiometric SiC at a few μm from the surface. The properties of a CVD SiC monofilament, summarized in Table 6.6, are superior to those of Nicalon fiber.

Table 6.6 *Properties of CVD SiC monofilament.*

Composition	Diameter (μm)	Density (g cm^{-3})	Tensile strength (MPa)	Young's modulus (GPa)
β–SiC	140	3.3	3500	430

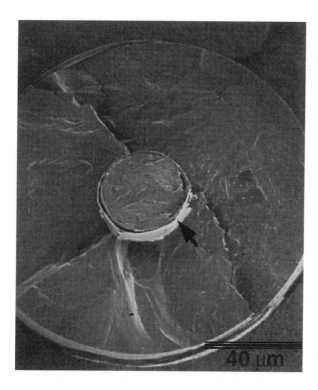

Figure 6.23 Fracture surface of an SCS-6 fiber after high frequency fatigue. Note the flaking in the pyrolytic coating on the carbon core (indicated by arrow) (courtesy of N. Chawla).

Recall that the properties of Nicalon start degrading above about 600°C because of the thermodynamic instability of composition and microstructure. The superior properties of CVD SiC fibers stem from the stoichiometric SiC, although the composite nature of these fibers can lead to problems. Figure 6.23 shows the fracture surface of an SCS-6 fiber after high frequency fatigue. Note the flaking in the pyrolytic coating on the carbon core (indicated by arrow). Large residual stresses and structural defects can be present at various interfaces in an SCS-6 fiber (Lara-Curzio and Sternstein, 1993; Kun *et al.*, 1996). These residual stresses result from the high processing temperature and the large anisotropy in the pyrolytic graphite layer.

The structure of Nicalon fiber has been studied by many researchers. Figure 6.24 shows a high resolution transmission electron micrograph of Nicalon-type SiC produced in the laboratory, indicating the amorphous nature of the SiC. Figures 6.25 (a) and (b) show the atomic force microscopy views of the surface of Nicalon and Hi-Nicalon fiber, respectively (Chawla *et al.*, 1995). Note the nodular surface in both cases. The commercial variety of Nicalon has an amorphous structure as shown in Fig. 6.24. A noncommercial variety, processed under special conditions, showed a microcrystalline structure (SiC grain radius of 1.7 nm) (Laffon *et al.*, 1989). The microstructural analysis shows that both the fibers contain, in addition to SiC, SiO_2 and free carbon. The density of the fiber is about $2.6 \, g/cm^3$ which is low compared to that of pure β-SiC. This is not sur-

Figure 6.24   High resolution transmission electron micrograph of a Nicalon produced in the laboratory, indicating the amorphous nature of this fiber (courtesy of K. Okaumra).

prising in view of the fact that the composition is a mixture of SiC, SiO_2 and C. A comparison of Nicalon SiC fiber with CVD SiC fiber shows that the CVD fiber is superior in properties (DiCarlo, 1994; DiCarlo *et al.*, 1995). It turns out that the continuous phase in these commercial *SiC*-type fibers (Nicalon and Tyranno) is amorphous, and it is this continuous phase that controls the ultimate properties of these fibers. The elastic moduli, for example, of these fibers are much lower than that of the pure, crystalline β-SiC. These fibers also have lower density than the theoretical density of crystalline β-SiC and, more importantly, they are not stable at high temperatures (Lipowitz *et al.*, 1994).

High temperature stability of these nonoxide fibers in air is another critical problem. Thermal stability of ceramic fibers derived from polymeric precursors is of special concern mainly because, as mentioned above, they frequently have undesirable phases present in them. Polycarbosilane-derived SiC fibers, such as Nicalon or Tyranno, involve a thermal oxidation curing process as described above and can contain as much as 10 wt% oxygen (Okamura and Seguchi, 1992). Such fibers decompose at temperatures above 1200°C in a nitrogen or argon atmosphere with SiO and CO gas evolution:

$$SiC_xO_y \rightarrow SiC(s) + SiO(g) + CO\ (g)$$

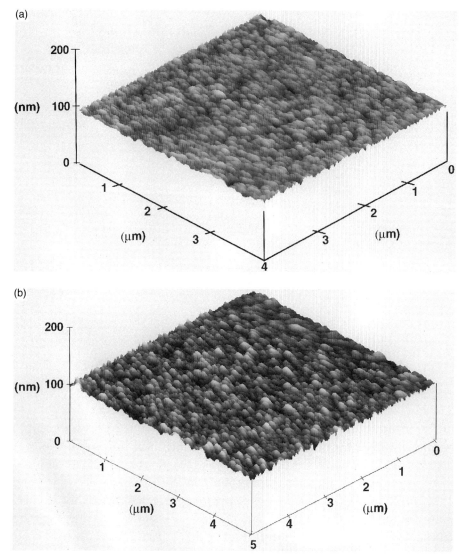

Figure 6.25 Topographic views of the surface of (a) Nicalon and (b) Hi-Nicalon fiber (courtesy of N. Chawla). Note the nodular surface in both cases.

This reaction is accompanied by a change in morphology and crystal structure of the fiber and a loss of tensile strength. In order to control the oxygen content of such fibers, Okamura *et al.* (1992) adopted a curing process using radiation–chemical reactions and produced oxygen-controlled SiC fibers. For example, by radiation curing in an oxygen atmosphere, they could make SiC fibers with a gradient in oxygen content. By electron radiation curing polycarbosilane in vacuum or helium, SiC fibers with less oxygen were prepared. These low oxygen fibers showed high strength and Young's modulus to temper-

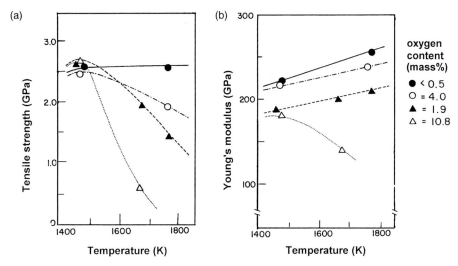

Figure 6.26   Tensile strength and modulus of Nicalon type fibers as a function of temperature and oxygen content (after Okamura *et al.*, 1993).

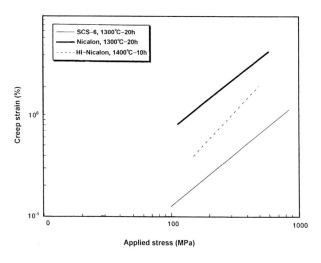

Figure 6.27   Comparison of the creep behavior of Nicalon, Hi-Nicalon and CVD SiC fiber. Note the superior performance of the CVD fiber, which is attributable to stoichiometric β-SiC (after Dicarlo, 1994)

atures as high as 1500°C and, at a given temperature, the strength and modulus increased with decreasing oxygen content. Figure 6.26 summarizes their results on tensile strength and modulus as a function of temperature and oxygen content. A comparison of the creep behavior of Nicalon, Hi-Nicalon, and CVD SiC fiber is shown in Fig. 6.27. Note the superior performance of the CVD fiber, which is attributable to stoichiometric β-SiC.

Other nonoxide fibers

Besides silicon carbide based ceramic fibers, there are other promising ceramic fibers, e.g. silicon nitride, boron carbide, boron nitride, etc.

Silicon nitride (Si_3N_4) fibers can be prepared by reactive chemical vapor deposition (CVD) using volatile silicon compounds. The reactants are generally $SiCl_4$ and NH_3. Si_3N_4 is deposited on a carbon or tungsten substrate. Again, as in other CVD processes, the resultant fiber has good properties but the diameter is very large and the fiber is expensive. In the polymer route, organosilzane polymers with methyl groups on Si and N have been used as silicon nitride precursors.

Polysilazanes have the right elements to give the silicon nitride, however, they are very reactive to water and tend to incorporate oxygen to form Si–O bonds easily. Polycarbosilanes can be nitrided between 500 and 800°C in a nitrogenous atmosphere such as ammonia. However, such carbon-containing, silicon–nitrogen precursors on pyrolysis give silicon carbide as well as silicon nitride, i.e. the resulting fiber is not an SiC-free silicon nitride fiber. Wills *et al.* (1983) have discussed the mechanisms involved in the conversion of various organometallic compounds into ceramics.

A boron nitride fiber can be very competitive commercially with carbon fiber. It has about the same density ($2.2\,g/cm^3$) as the carbon fiber, but has greater oxidation resistance and excellent dielectric properties. A method of converting boric oxide precursor fibers into boron nitride fibers has been developed (Economy and Anderson, 1967). Melt spun boric oxide precursor fiber is nitrided with ammonia according to the following reaction:

$$B_2O_3 + 2NH_3 \rightarrow 2BN + 3H_2O$$

This is followed by a high temperature treatment that removes the traces of oxides and stabilizes the product (Smith, 1977).

The preceramic method has been applied to make BN fibers. The trick again is to synthesize a preceramic fibrous precursor having the same elements as the ones desired in the final ceramic fiber. Many derivatives of poly(aminoborazinyls) can be used as precursors for BN fibers (Fazen *et al.*, 1990; Paciorek *et al.*, 1986; Narula *et al.*, 1990) but their spinnability is rather poor. Kimura *et al.* (1994) made a hexagonal boron nitride fiber via a preceramic method based on a thermoplastic derivative of poly(aminoborazinyl) as the preceramic fiber. The preceramic polymer was made by thermal condensation of B,B,B-tri(methylamino)borazine (MAB) and was melt spun into the preceramic fiber. The preceramic fiber is structurally similar to melt-spinnable carbon pitch. The precursor fiber becomes slightly hydrolyzed at the surface during the spinning process, and thus becomes infusible. This allows the fiber to be sintered at temperatures up to 1000°C in ammonia gas flow without resorting to a thermosetting step. At this stage there is some contamination with carbon from the organic residue and the fiber appears black. This is followed by pyrolysis up to 1800°C in flowing nitrogen gas, with the color changing to brown at 1200°C and white above 1400°C.

Boron carbide is also a very light and strong material. It can be prepared by reacting carbon yarn with BCl_3 and H_2 at high temperatures, i.e. a CVD process (Economy and Lin, 1977). The chemical reaction involved is

$$4BCl_3 + 6H_2 + C_{fibers} \rightarrow B_4C_{fibers} + 12 \text{ HCl}$$

The reaction actually occurs in two steps, namely

$$2BCl_3 + 3H_2 \rightarrow 2B + 6HCl$$

$$4B + C \rightarrow B_4C$$

Much like other CVD processes described earlier, the gaseous mixture BCl_3, H_2, and argon (diluent) enter at one end of a furnace, react in the hot zone, and the reaction products exit at the other end. The second step in the reaction above is the rate controlling step as the reaction of B and C is slowed by the formation of a B_4C layer.

Another candidate material for high temperature fiber is titanium diboride. It has a melting point of around 3000°C. Diefendorf and Mazlout (1994) used a gas mixture of titanium tetrachloride, boron trichloride, hydrogen, and hydrochloride to make titanium diboride fibers by chemical vapor deposition (CVD) in a cold wall reactor at atmospheric pressure.

We make a mention of phosphate fibers, or more appropriately polyphosphate fibers, that were developed by Monsanto Co. as a replacement for asbestos fibers (Griffith, 1995). Basically, the silicon atoms of asbestos were replaced by phosphorus atoms. These polyphosphate fibers were supposed to be a kind of 'safe' asbestos fiber. However, Monsanto did not commercialize these fibers because of a possible vulnerability to lawsuits. The safety feature of polyphosphate fibers stems from the fact that they are composed of components that are constituents of the human body and are thus biodegradable. The polyphosphate fibers developed by Monsanto were similar to asbestos fibers in terms of their insulating and nonflammable characteristics.

Calcium polyphosphate fibers were grown from a water soluble melt. The raw materials used were lime (CaO) and phosphoric acid (H_3PO_4) to give

$$nCaO + 2n\,H_3PO_4 \rightarrow [Ca(PO_3)_2]_n + 2nH_2O$$

Calcium polyphosphate fibers were grown as dendritic spherulites. Sodium calcium polyphosphates were grown as simple crystals and then milled.

Finally, there are potassium titanate fibers that are produced by heating titanium dioxide and potassium carbonate in a furnace to produce whisker-like short fibers, about 0.5 μm in diameter and 50 μm or more in length. Potassium titanate fibers are quite strong and suitable for reinforcement of polymers. At one time, Du Pont made them in the US. Their production in the US was discontinued because of safety reasons. However, these fibers are available commercially in Japan.

6.4.3   Boron fibers

Boron fibers are commercially made by chemical vapor deposition of boron on a substrate. Thus, a boron fiber, similar to the SiC CVD fiber described above, is itself a composite fiber. Because of the high temperatures required for this deposition process, the choice of substrate material which forms the core of the finished boron fiber is limited. Generally, a fine tungsten wire is used for this purpose, but a carbon substrate can also be used. The first boron fibers were obtained by Weintraub (1911) by reducing a boron halide with hydrogen on a hot wire substrate. These boron fibers were not of high strength. Talley (1959) used the process of halide reduction to obtain amorphous boron fibers of high strength. This work opened the way for strong but light boron fibers as possible structural components.

Fabrication of boron fibers

Commercially, boron fibers are obtained by chemical vapor deposition on a substrate. The process involves reduction of boron trichloride by hydrogen gas according to the following reaction:

$$2\ BC\ell_3 + 3\ H_2 \rightarrow 2\ B + 6\ HC\ell$$

In this process of halide reduction, the temperatures involved are very high, so one needs a refractory material, e.g. high melting point metal such as tungsten, as a substrate. Tungsten being a very heavy metal (its density is $19.3\ \mathrm{g\,cm^{-3}}$), its use as a substrate results in a boron fiber having a density higher than that of elemental boron. This process, however, gives boron fibers of high quality.

Boron fiber is produced by chemical vapor deposition on a tungsten wire in a manner similar to that described for the CVD silicon carbide fibers in Fig. 6.17 or Fig. 6.19. A fine tungsten wire (10–12 μm diameter) is pulled into a reaction chamber at one end through a mercury seal and out at the other end through another mercury seal. The mercury seals act as electrical contacts for resistance heating of the substrate wire when gases ($BC\ell_3 + H_2$) pass through the reaction chamber where they react on the incandescent tungsten wire substrate. $BC\ell_3$ is an expensive chemical and only about 10% of it is converted into boron in this reaction. Thus, efficient recovery of the unused $BC\ell_3$ can result in a considerable lowering of the boron filament cost.

There is a critical temperature for obtaining a boron fiber with optimum properties and structure (van Maaren *et al.*, 1975). The desirable microcrystalline form of boron with a grain size of about 2–3 nm occurs below this critical temperature, while above this temperature crystalline forms of boron also occur. Crystalline boron does not have very good mechanical properties. With the substrate wire stationary in the reactor, the critical temperature is about 1000°C. In a system where the wire is moving, the critical temperature is higher and increases with the speed of the wire. Various combinations of wire temperature and wire drawing speed can be used to produce a certain diameter of boron fiber. Boron is deposited in an amorphous state and the more rapidly the wire is drawn out from the reactor, the higher is the allowed temperature. Of course, a higher wire drawing speed also results in an increase in production rate and lower costs.

Boron deposition on a carbon monofilament core involves precoating a carbon substrate (~35 μm diameter) with a 1–2 μm layer of pyrolytic carbon. The pyrolitic carbon coating is applied by exposing the carbon core to a mixture of methane, argon and hydrogen at about 2500°C. The pyrolytic carbon coating accommodates the growth strains that result during B deposition (Krukonis, 1977). The reactor assembly is slightly different from that used to produce boron on a tungsten substrate, as pyrolytic carbon is applied online.

Structure and morphology

The structure and morphology of boron fibers depend on the conditions of deposition: temperature, composition of gases, gas dynamics, etc. Temperature gradients and trace concentrations of impurity elements inevitably cause process irregularities. Even greater irregularities are caused by fluctuations in electric power, instability in gas flow or any other operator-induced variables. These irregularities can lead to structural defects and morphological irregularities which lower the mechanical properties of boron fiber.

Structure

Elemental boron can have different crystalline polymorphs, depending on the conditions of deposition. On crystallization from the melt or chemical vapor deposition above 1300°C, one gets β-rhombohedral. At temperatures below 1300°C, the most commonly observed structure is α-rhombohedral.

Boron fibers produced by the CVD method described above have a microcrystalline structure that is generally called 'amorphous'. This designation is based on the characteristic X-ray diffraction pattern produced by the filament in the Debye–Scherrer method, i.e. large and diffuse halos, typical of an amorphous material (Vega-Boggio and Vingsbo, 1978). Electron diffraction studies show this boron fiber to be nanocrystalline with an average grain size of about 2 nm (Krukonis, 1977). It would thus appear that amorphous boron is really a

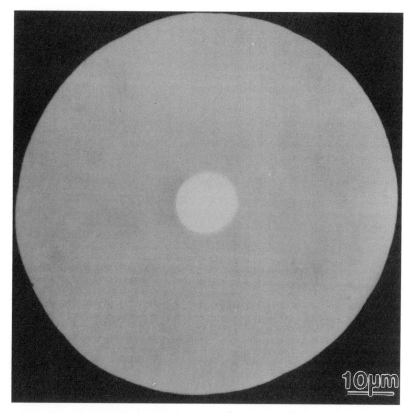

Figure 6.28   Cross-section of a boron fiber.

nanocrystalline β-rhombohedral structure. Should any large crystals or groups of crystals be formed, they will constitute an imperfection in the fiber, and should be avoided. Such imperfections generally result from exceeding the critical temperature of deposition or due to the presence of impurities in the gases.

When boron fibers are made by deposition on a tungsten substrate, as is generally the case, then, depending on the temperature conditions during deposition, the core may consist of, in addition to W, a series of compounds, such as W_2B, WB, W_2B_5 and WB_4. A cross-section of such a boron fiber is shown in Fig. 6.28, while Fig. 6.29 shows schematically the various subparts of a boron fiber cross-section. The tungsten boride phases are formed by diffusion of B into W. Generally the fiber core consists only of WB_4 and W_2B_5. On prolonged heating, the core may completely be converted to WB_4. As boron diffuses into the tungsten substrate to form borides, the core expands from its original 12.5 μm (original W wire diameter) to 17.5 μm. When boron fibers are used as reinforcement for metallic matrices such aluminum or titanium for high temperature use, a coating of SiC or boron carbide is applied to prevent chemical reaction between the matrix and the boron fiber.

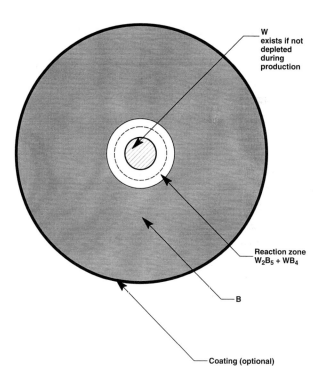

W
exists if not
depleted
during
production

Figure 6.29 Various sub-parts of a boron fiber cross-section.

Reaction zone
W$_2$B$_5$ + WB$_4$

B

Coating (optional)

Morphology

The surface of boron fiber shows a *corn-cob* structure consisting of nodules separated by boundaries, Fig. 6.30. The nodule size varies during the course of fabrication and has its origin in the very nature of the CVD process. Almost any fiber-making process via CVD on a substrate results in such a nodular surface morphology to some extent. This is particularly exacerbated by a tungsten wire substrate because it has longitudinal ridge markings that result from the wire drawing process. These longitudinal ridges provide preferential nucleation sites for the boron nodules, which start as individual nuclei on the substrate and then grow outward in a conical fashion until a filament diameter of 80–90 μm is reached, above which the nodules seem to decrease in size. Occasionally, new cones may nucleate in the material, but they always originate at an interface with a foreign particle or inclusion.

Residual stresses in boron fiber

Boron fibers, like any CVD fiber, have inherent residual stresses which originate in the process of chemical vapor deposition. Growth stresses in the nodules of boron, stresses induced due to diffusion of boron into the W core, and stresses generated due to the difference in the coefficient of expansion of the deposited boron and the tungsten boride core, all contribute to the residual stresses, and thus can have a considerable influence on the fiber mechanical properties.

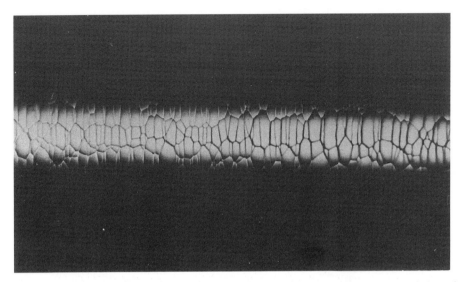

Figure 6.30   Surface of a boron fiber showing a *corn–cob* structure consisting of nodules separated by boundaries (courtesy of Phillips Electronics).

Boron and silicon carbide fibers are produced by CVD on a tungsten wire substrate. B(W) has radial cracks due to the fact that the boron fiber consists of B, W_xB_y, and almost no W at the end of the process. Volume changes involved in the formation of compounds are constrained by the surrounding sheaths of unreacted material leading to residual stresses (Adler and Hammond, 1969; Vega-Boggio and Vingsbo, 1976a, 1976b; Layden, 1973). SiC(W), on the other hand, consists of W and SiC, with little or no reaction. One has to contend with a very small amount of differential thermal mismatch, and no compound formation occurs unless exposed to very high temperatures. Radial cracks form in boron fiber from the core outward. Axial extension is considerable, while tangential extension is negligible. The final quench results in a compressive surface layer and the radial crack stops at the boundary between tensile and compressive stress where the tensile stress is zero. A schematic of the residual stress pattern across the transverse section of a boron fiber is shown in Fig. 6.31. The compressive stresses on the fiber surface are due to the quenching action involved in pulling the fiber from the chamber.

As pointed out above, boron fiber is also a composite fiber like any other CVD-type fiber. During fracture, cracks frequently originate at pre-existing defects located at the boron/core interface or at the surface. Commonly, a radial crack originates at the boron/core interface and leads to a brittle fracture.

Properties of boron fibers

Due to the composite nature of boron fibers, complex internal stresses and defects such as voids and structural discontinuities result from the presence of a

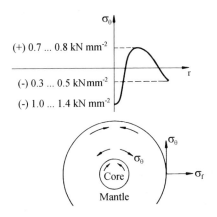

Figure 6.31 The residual stress pattern across the transverse section of a boron fiber (after Vega-Boggio and Vingsbo, 1978).

core and the deposition process. Thus, one would not expect boron fiber strength to equal the intrinsic strength of boron. The average tensile strength of boron fibers is 3–4 GPa while Young's modulus is between 380 and 400 GPa (Wawner, 1967).

The intrinsic strength of boron can be estimated in a flexure test. Assuming that in a flexure test the core and interface are near the neutral axis, critical tensile stresses would not develop at the core or interface. Flexure tests on boron fibers lightly etched to remove any surface defects yield a strength of 14 GPa (DiCarlo, 1985). Without etching, the strength is about half this value.

Table 6.7 summarizes the properties of boron. The high temperature treatment, indicated in this table, improves the fiber properties by putting a permanent axial compressive strain in the sheath. Commercially produced 142 μm

Table 6.7 *Properties of boron fibers (after Dicarlo, 1985).*[a]

Diameter (μm)	Treatment	Average strength (GPa)	COV[b] (%)	Relative fracture energy
142	As-produced	3.8	10	1.0
406	As-produced	2.1	14	0.3
382	Chemical polish	4.6	4	1.4
382	Heat treatment plus polish	5.7	4	2.2

Notes:
[a] Gage length=25 mm.
[b] Coefficient of variation=standard deviation/average value.

diameter boron fiber shows tensile strengths <3.8 GPa. The tensile strength and fracture energy values of the as-received and some limited-production run larger diameter fibers show improvement after chemical polishing (DiCarlo, 1985). Fibers showing strengths above 4 GPa have their fracture controlled by the tungsten boride core while strengths of 4 GPa or less are controlled by fiber surface flaws. Chemical polishing improves the strength because it removes some of the defects from the surface region.

Boron has a density of 2.34 g cm^{-3} (about 15% less than that of Al). Boron fibers with tungsten core have a density of 2.6 g cm^{-3} for a fiber of 100 μm diameter. Its melting point is 2040°C and it has a thermal expansion coefficient of 4.86×10^{-6} K^{-1} up to 315°C.

As mentioned above, commercially produced boron fibers have a large diameter, 142 μm. This makes them rather inflexible (see Chapter 2 for flexibility). Wallenberger and Nordine (1992) have made smaller diameter boron fibers, <25 μm, by laser assisted chemical vapor deposition (LCVD). This process uses the focal point of a laser beam rather than a heated tungsten substrate as the heat source to decompose the gaseous reactants at pressures >1 bar. The smallest fiber diameter achievable in this process is limited by the size of the laser focal spot size which in turn is limited by diffraction. The process involves continual adjustment of the hot focal point of the laser to coincide with the tip of the growing fiber and to support the continuous growth of the fiber. According to these authors it is possible to use the LCVD process to make other small diameter inorganic fibers such as silicon carbide.

6.5   Ceramic whiskers

Whiskers are monocrystalline, short fibers with extremely high strength. Strength levels approaching the theoretically expected values have been measured in many whiskers (Brenner, 1958, 1962). α-alumina whiskers showed very high strength to very high temperatures (Brenner, 1962). This high strength, approaching the theoretical strength, has its origin in the absence of crystalline imperfections such as dislocations. Being monocrystalline, there are no grain boundaries either. Typically, whiskers have a diameter of a few μm and a length of a few mm. Thus, their aspect ratio (length/diameter) can be very high, varying from 50 to 10 000. This makes it attractive to use these high strength whiskers as reinforcements. Whiskers do not have uniform dimensions or properties. This is perhaps their greatest disadvantage, i.e. the spread in properties is extremely large. It should also be recognized that preservation of a defect-free surface during processing of a composite is a very difficult task, i.e. handling and alignment of whiskers in a matrix to produce a composite are rather difficult.

Whiskers are normally obtained by vapor phase growth. Early in the 1970s, a new process was developed, starting from rice hulls, to produce SiC particles and whiskers (Lee and Cutler, 1975; Hollar and Kim, 1991). The SiC particles produced by this process are of a fine size. Rice hulls are a waste by-product of rice milling. For each 100 kg of rice milled, about 20 kg of rice hull is produced. Rice hulls contain cellulose, silica and other organic and inorganic materials. Silica from soil is dissolved and transported in the plant as monosilicic acid. This is deposited in the cellulosic structure by liquid evaporation. It turns out that most of the silica ends up in the hull. It is the intimate mixture of silica within the cellulose that gives the near ideal amounts of silica and carbon for silicon carbide production. Raw rice hulls are heated in the absence of oxygen at about 700°C to drive out the volatile compounds. This is called coking. Coked rice hulls, containing about equal amounts of SiO_2 and free C, are heated in an inert or reducing atmosphere (flowing N_2 or NH_3 gas) at a temperature between 1500 and 1600°C for about 1 hour to form silicon carbide in accordance with the following reaction

$$3C + SiO_2 \rightarrow SiC + 2CO$$

Figure 6.32 shows a schematic of the process. When the above reaction is over, the residue is heated to 800°C to remove any free C. Generally, both particles and whiskers are produced together with some excess free carbon. A wet process is used to separate the particles and the whiskers. Typically, the average aspect ratio of the as-produced whiskers is 75.

A vapor–liquid–solid (VLS) technique (Petrovic et al., 1985; Milewski et al., 1985) can be used to grow exceptionally strong and stiff silicon carbide whiskers. The name VLS comes from the fact that the process uses vapor feed gases, a liquid catalyst, and solid crystalline whiskers are the end product. Figure 6.33 shows the process schematically. The catalyst forms a liquid solution interface with the growing crystalline phase while elements are fed from the vapor phase through the liquid–vapor interface. Whisker growth takes place by precipitation from the supersaturated liquid at the solid/liquid interface. The catalyst must take in solution the atomic species of the whisker to be grown. For SiC whiskers, transition metals and iron alloys meet this requirement. One of the major drawbacks of whiskers is the extreme variability in their properties. For example, the properties of SiC whiskers grown by the VLS technique showed average tensile strength and modulus of 8.4 GPa and 581 GPa, respectively. However, the tensile strength values ranged from 1.7 to 23.7 GPa in 40 tests. Whisker lengths were about 10 mm and the equivalent circular diameter averaged 5.9 μm.

6.6   Applications

Refractory or ceramic fibers have low thermal conductivity, low heat capacity and high chemical resistance. These characteristics make them attractive as

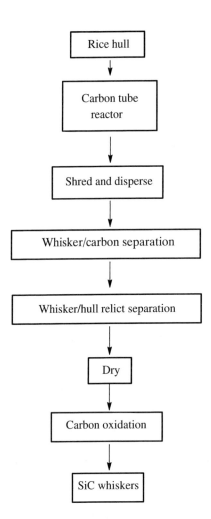

Figure 6.32 Schematic of the rice hull process of making silicon carbide whiskers.

insulation materials in metal processing industries, where they are frequently used as insulation in furnaces. The characteristics of chemical inertness to molten metals and natural gas and low thermal capacity are especially useful. Whenever a furnace is cooled, e.g. during shut down or as part of an operating cycle, much of the stored heat in the insulating lining is lost. A low thermal capacity lining will reduce such heat losses. It will also allow rapid cooling and heating. Aluminosilicate fibers, e.g. Fiberfrax, Nextel, etc. have very low thermal conductivity and thus are excellent for insulating purposes. Applications include common ones such as various high temperature furnace seals and insulation, and special ones such as insulating catalytic converters and mufflers on vehicles, copying machine components, door seals in stoves, chemical tanks and cylinders, etc. Such fibers are available in a variety of different product forms such as blanket, board, paper, woven fabric, tape, yarn, braided sleeves, rope, etc. Braided thermocouple insulation is a very common application for high tem-

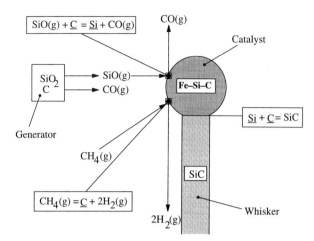

Figure 6.33 Schematic of the VLS process of making silicon carbide whiskers (after Petrovic *et al.*, 1985).

$$SiO(g) + \underline{2C} = SiC(s) + CO(g)$$

perature measurement. Another application of such fibers is in the making of high temperature filter bags. Special mention should be made of a sprayable form of fiber, coated with a binder, that can be sprayed on to, say, a metal surface, forming an interlocked network of short fibers and highly efficient insulating air pockets. The spray process is very fast and is used for refits and repairs in the chemical, metallurgical and ceramic industry. Such alumina–silica insulating fibers are commonly used up to service temperatures of 1250°C while alumina, mullite and zirconia can extend usable temperatures to 1600°C. Mention should be made of insulation boards made of mullite fibers+fillers+inorganic binders by Rath Co., called Altra Boards, with enhanced thermal shock resistance for kiln furnace linings up to 1800°C (Anon., 1993). This product optimizes divergent properties of thermal shrinkage and thermal shock resistance.

Use of ceramic fiber burners with natural gas can result in reduced emissions of nitrogen oxide. Ceramic fiber burners can lead to a high radiative heat transfer and thus a lower flue gas temperature, which in turn results in lower nitrogen monoxide emission. The nitrogen oxide formation depends on the flue gas temperature in a more or less exponential manner. Thus, a reduced flue gas temperature results in a reduced NO_x production. Ceramic fiber burners can thus lead to less NO_x than conventional burners.

High stiffness ceramic fibers such as alumina, alumina–silica, silicon carbide, boron, etc. are used as reinforcement fibers for polymeric, metallic, and ceramic matrix composites (Chawla, 1987). Silicon carbide whisker reinforced alumina composites are used as high speed cutting tools (Chawla, 1993).

Chapter 7

Glass fibers

The term glass or a glassy material represents a rather large family of materials with the common characteristic that their structure is noncrystalline. Thus, rigorously speaking, one can produce a glassy material from a polymer, metal or ceramic. An amorphous structure is fairly common in polymeric materials. It is less so in metals, although metallic glass, generally in the form a ribbon, can be produced by rapid solidification, i.e. by not giving enough time for crystallization to occur. In this chapter we describe silica-based inorganic glasses because of their great commercial importance, as a reinforcement fiber for polymer matrix composites and as an optical fiber for communications. Communication via optical glass fibers is a well established field. Crude optical glass fiber bundles were used to examine the insides of the human body as far back as 1960. Since then tremendous progress has been made in making ultra pure, controlled composition fibers with very low optical attenuation. It is estimated that the total worldwide shipment of optical fibers is over US$ 5 billion per year. Before we describe the processing, structure, and properties of glass fiber, it would be appropriate to digress slightly and describe for the uninitiated, albeit very briefly, the basic physics behind the process of communication via optical glass fibers.

7.1   Basic physics of optical communication

Optical glass fiber has many desirable characteristics for communication such as:

- large bandwidth over great distances;
- protection against electrical interference and crosstalk;

- galvanically protected signal transmission;
- safety and reliability.

By far the most important and simple phenomenon that is made use of in optical wave guides is *refraction* of light. When a ray of light propagates from one medium to another, it undergoes refraction at the interface such that the ratio of the sine of the incident angle and the sine of the refracted angle equals the inverse ratio of the relative refractive index of the two media. Thus,

$$\sin \theta_1 / \sin \theta_2 = n_2 / n_1$$

where θ_1 is the incident angle, θ_2 is the refracted angle, and n_1 and n_2 are the refractive indexes of medium 1 and 2, respectively. This is called Snell's law, and it gives the exit angle as

$$\theta_1 = \sin^{-1}(n_2 \sin \theta_2 / n_1)$$

It is easy to see from this expression that for some critical angle and a given refractive index ratio, θ_2 will equal 90° and $\sin \theta_2$ will be unity. For all angles greater than the critical angle, θ_c, the incident light will be totally reflected, and we can write

$$\sin \theta_{2max} = 1$$
$$\sin \theta_c = n_2 / n_1$$
$$\theta_c = \sin^{-1}(n_2 / n_1)$$

For $n_1 > n_2$, we shall have a critical angle, $\theta_1 = \theta_c$ called the critical angle for total internal reflection such that no refracted ray comes out, i.e. total internal refraction occurs. This condition of total internal reflection is the key for the light propagation through a glass fiber and we satisfy this condition when the incident angle equals or exceeds the critical angle, θ_c. If we extend this idea to a situation where a ray propagates from a dense medium through successive layers of rarer medium, we can have the ray bend. In an extreme case of a medium having a continuously varying refractive index, the ray will bend and suffer *total internal reflection*, i.e. the light will be trapped in such a medium. It is this phenomenon of light trapping that is made use of in optical fiber for light transmission. Fiberoptic glass, thus, consists of a core that conducts the light and a cladding that has lower refractive index than the core. This enables a light beam to undergo a total internal reflection at the core/cladding interface, that is one has a light-guiding effect. Figure 7.1 shows this effect schematically. The reader should appreciate the key point here, namely that the light-conducting glass core should cause as little attenuation of the light signal as possible during its

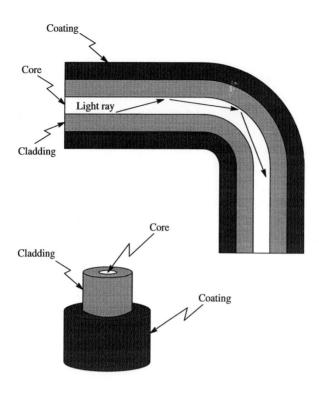

Figure 7.1 A light beam undergoing total internal reflection at the core/cladding interface. The bottom part shows a perspective view consisting of the light conducting core, the cladding that has a lower refractive index than the core, and the protective coating.

transmission over long distances. It is convenient to measure the optical attenuation of the glass fiber in decibels per kilometer (dB/km). The symbol dB is used indicate the logarithmic scale of decibels. The decibel scale is preferred when a quantity can vary over several orders of magnitude. Thus, we can express the optical power ratio or an index of attenuation of a plane wave A, more conveniently in the decibel scale as

$$A \text{ (dB)} = 10 \log_{10} I_0/I$$

where I_0 is the initial energy and I is terminal energy. Up until 1970, optical glass fibers had high losses, ~1000 dB/km, and were used mainly for short distance imaging for medical purposes. In the 1970s low loss glass fibers (as low as 0.2 dB/km) became available and the era of optical communications began. Let us see what this 0.2 dB/km translates into in terms of attenuation of the initial energy. Using the above expression, we can write

$$0.2 = 10 \log_{10} I_0/I$$

so that

$$\log_{10} I_0/I = 0.02$$

and

$$I_0/I = 1.05$$

giving

$$I = 0.95\, I_0$$

This is a pretty low index of attenuation!

For short range optical links, polymeric fibers made of a polystyrene or poly methyl methacrylate core fiber and low refractive index polymer coating can be used. The best reported attenuation figure for these is 200 dB/km. It should be recognized that the capacity of lasers to generate strong and intense optical signals has also been helpful in the great progress made in the field of communications via optical fibers. Nevertheless, the so-called photonics revolution could not have come about without the development of optical glass fibers that can transmit light efficiently over long distances. These optical fibers have transformed telecommunication and data networks.

7.2   **Fabrication**

We first describe the characteristic drawability feature of glasses which stems from their viscoelastic nature and then describe the fabrication processes for the glass fiber for insulation and reinforcement of polymers and for optical communication purposes, both of which exploit the ability of glass to be drawn into a very small diameter fiber.

7.2.1   Viscoelastic behavior of glasses

Viscoelasticity is the phenomenon of time-dependent strain. Often, it is also referred to as anelasticity. Glassy materials, above the glass transition temperature T_g show Newtonian viscosity, i.e. the stress is proportional to the strain rate. This property is exploited in the drawing of fiber and sheet forms. We can write, in terms of normal stresses and strains,

$$\sigma = \eta \dot{\epsilon} \qquad\qquad\qquad (7.1)$$

where σ is the stress, η is the viscosity, and $\dot{\epsilon}$ is the strain rate. The deformation proceeds at a constant volume, i.e. an increase in length is accompanied by a decrease in the cross-sectional area. If an incremental increase in length is $d\ell$, and the corresponding decrease in cross-section is $-dA$, then

$$\mathrm{d}\ell/\ell=\mathrm{d}\epsilon=-\mathrm{d}A/A$$

where ℓ and A are the instantaneous values of length and cross-sectional area, respectively. We introduce the strain rate by dividing throughout by incremental time, $\mathrm{d}t$. Thus,

$$\dot{\epsilon}=\mathrm{d}\epsilon/\mathrm{d}t=-(\mathrm{d}A/A)(1/\mathrm{d}t)=-(\mathrm{d}A/\mathrm{d}t)(1/A)$$
$$=-\dot{A}/A$$
$$A\dot{\epsilon}=-\dot{A} \tag{7.2}$$

If F is the applied load, then from Eq. (7.1) we have

$$F/A=\sigma=\eta\dot{\epsilon}$$

or, using Eq. (7.2), we can write

$$F=\eta\dot{\epsilon}A=-\eta\dot{A} \tag{7.3}$$

Equation (7.3) states that the rate at which the fiber (or sheet) becomes thinner is proportional to the applied force, *not the applied stress*. This means that thinner and thicker regions suffer cross-sectional reduction at an equal rate. This expression also informs us that, for a given load, as the viscosity η increases, the axial strain rate, $\dot{\epsilon}$, decreases. This has very important implications in fiber drawing. In order to understand these implications, we need to examine the temperature dependence of viscosity. The temperature dependence of viscosity is given by the Eyring equation

$$\eta=(hN/V_{\mathrm{m}})\exp{(Q/RT)}=A\exp{(Q/RT)}$$

where h is Planck's constant, N is Avogadro's number, V_{m} is the gram molecular volume of the liquid, Q is the molar activation energy, A is a pre-exponential constant and R is the universal gas constant. The Eyring equation indicates that the viscosity varies inversely in an exponential manner with the temperature. As thinner regions of the fiber cool more rapidly, the viscosity of such regions increases resulting in a decrease in the rate at which the thin region gets thinner. This allows the thicker regions to catch up and the fiber (or the film) extends uniformly without forming a *necked* region. This characteristic, due to Newtonian behavior, is exploited in the manufacture of glass fiber involving extension, through platinum orifices, of molten glass as explained below.

7.2.2 Glass fiber for reinforcement

Reinforcement of polymer matrix materials is one of the largest markets for glass fiber. A schematic of the conventional fabrication procedure for glass fibers is shown in Fig. 7.2a while a picture of glass fibers being drawn out of holes in a bushing is shown in Fig. 7.2b. The raw materials are heated in a hopper and the molten glass is fed into electrically heated, multiholed, platinum or platinum–rhodium bushings. Typically, each bushing contains 200 holes at its base. A constant head of molten glass is kept in the tank. The molten glass flows by gravity through these holes forming fine continuous filaments which are gathered together and passed around a fast rotating collet, followed by drawing at a speed of 1–2 km/min. Glass flows out of a platinum bushing (200–400 holes) at a rate given by:

$$q = khr^4/\nu\ell$$

where r is the radius, ℓ is the length, ν is the kinematic viscosity (i.e. dynamic viscosity η divided by density, ρ), and h is the hydrostatic pressure generated by the head of glass in the bushing. The fibers exiting the bushing are cooled, subjected to a surface treatment (an organic *size* such as a starch oil), and given a stretching treatment by the take-up spool.

Typically, the glass fibers have a diameter between 5 and 20 μm. A size, consisting of an aqueous polymer emulsion, is applied before winding on a drum. The final glass fiber diameter is a function of the bushing orifice diameter, viscosity, which is a function of composition and temperature, and the head of glass in the hopper. The viscosity is generally around 100 Pa s for good fiberization. The bushing diameter (1–2 mm), the head of molten glass, and the viscosity control the drawing rate of fibers. The glass fibers undergo a stretching treatment by the take-up spool. The take-up speed is 50–60 ms^{-1}, which is slightly higher than the drawing speed, and thus the as-wound glass fiber suffers a stretch.

Glass filaments are easily damaged by the introduction of surface defects. In order to minimize this and ease the handling of these fibers, the sizing treatment protects as well as binds the filaments into a strand. In almost any fiber drawing process from a melt, the melt viscosity is by far the most important process control parameter. If a single glass filament breaks, the glass will drip across other filaments and cause a breakdown of the whole drawing process. Filament breakage can occur because of undissolved particles in the glass, abrasion, devitrification, or any instability due to changes in viscosity/winding speed (Proctor, 1971).

A very important characteristic of glass fiber is its extreme flexibility, which of course stems from its fine diameter and low modulus. Textile processes of weaving, knitting, etc. can be used with fine, flexible glass fibers to produce

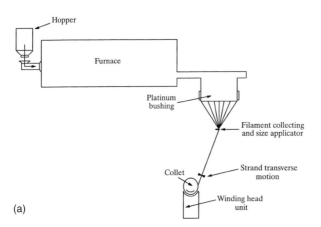

(a)

Figure 7.2   (a) Schematic of the conventional fabrication procedure for glass fibers (b) Glass fibers being drawn out of holes in a bushing (courtesy of Owen-Corning, Inc.).

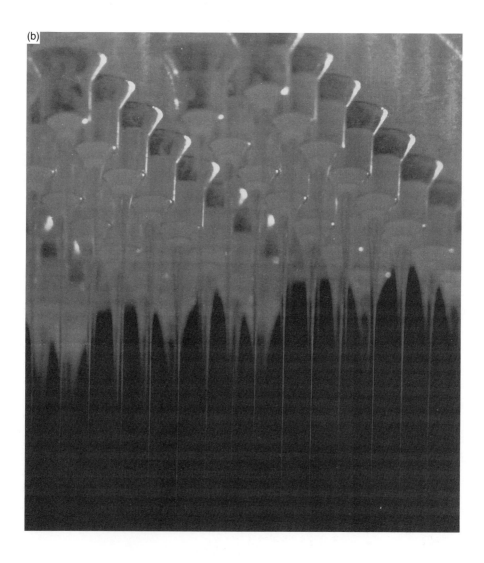

fabrics, tapes, sleeves, and ropes. E-glass fiber has a diameter between 3 and 20 μm and is commonly used for insulation purposes as well as for reinforcement of polymers. Frequently, glass fibers are chopped to obtain short, staple fibers. Glass wool, used extensively for insulating purposes, is made by a rotary or spinner process. A vessel containing molten glass is subjected to rapid rotation and glass fibers are spun from a series of rotating heads as a result of the centrifugal force. The emerging horizontal glass streams are impacted by blasts of combustion gases at high speed converting molten glass into short fibers. Binder is sprayed onto these fibers during their downward journey to a conveyor belt to give them a matted form for insulating material. Finally, the binder is cured in an oven and the glass wool is compressed by rollers to the desired thickness.

This characteristic of flexibility can be exploited in some rather innovative ways. For example, Sandtex Co. makes silica fiber from glass fiber by a thermal and chemical digestion process that converts a common glass fiber precursor to essentially pure silica ($>98\%$ SiO_2). The thermal and chemical leaching process removes most of the metallic oxides from the precursor glass fiber while maintaining the fibrous form. Thorough washing removes acidic residues and the application of proprietary surface finishes gives handleability and abrasion resistance. Figure 7.3a shows some forms in which silica glass fiber can be obtained by this process. A special glass fiber, trade name Miraflex (Owens–Corning) is shown in Fig. 7.3b. Note the different contrast in the two halves of the fiber cross-section. It is a bicomponent fiber and is made by fusing together two glass formulations having different coefficients of expansion. This structure gives the fiber natural, irregular twists.

Sol–gel processing of glass fiber

The conventional processing of glass fibers, as described above, involves direct melting of raw materials, followed by drawing of the molten glass through electrically heated platinum bushings. A major problem with any direct melting process is that the temperatures involved are very high. Quite frequently, the melt viscosity at reasonable temperatures is too high to allow proper homogenization. At times the oxides show large differences in volatility and direct melting results in large losses of some of the constituents. All these problems, as pointed out in Chapter 6, have led to processing of ceramics and glasses via a chemical route. This allows a high degree of homogeneity on a molecular scale and consequently high purity glasses or ceramics can be obtained. In particular, the sol–gel technique of making glass fibers has become important.

We give a brief description of the general principles of the sol–gel technique, with special emphasis on fiber making via this technique. Essentially the sol–gel route of making any glass or ceramic involves the formation of the appropriate glass or ceramic structure by chemical polymerization of suitable compounds in

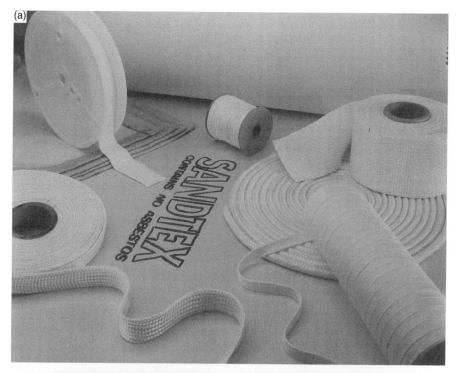

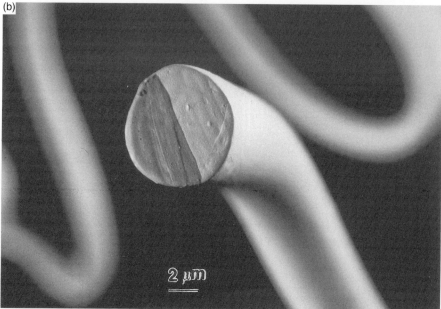

Figure 7.3   (a) Different forms of high silica (>98% SiO$_2$) glass fiber: such a fiber is obtained by a thermal and chemical digestion process that converts a common glass fiber precursor to essentially pure silica (courtesy of Sandtex Co.) (b) A bicomponent glass fiber, Miraflex (courtesy of Owens-Corning). Note the different contrast in the two halves of the cross-section.

the liquid state (sol) at low temperatures. A sol is a colloidal suspension in which the individual particles are so small that no sedimentation occurs. The particle size in a sol generally varies between 1 and 100 nm. Gel, on the other hand, is a suspension in which the liquid medium is viscous enough to behave more or less like a solid. A precursor material in the form of a gel is obtained from the sol. The unwanted residues (water, organic compounds, etc.) are removed from the gel by heating at temperatures much lower than those used in direct melting processes, and the desired glass or ceramic is obtained in an appropriate form (powder, film, fiber, etc.). As the solvent liquid is lost, the sol viscosity increases until it becomes rigid, i.e. it gels. The gel is the starting material for conversion into glass or ceramic. Figure 7.4a shows the sol–gel process flow diagram while Fig. 7.4b shows the microstructural changes that occur during the process. Most sol–gel processing involves the formation of metal oxides (ceramics or glasses) from metal alkoxides. A metal alkoxide has the chemical formula of $M(OR)_n$, where M is a metal or metalloid and R is an alkyl group such as CH_3, C_2H_5, etc., and n is the valence of the metal atom. Hydrolysis of metal alkoxides provides a low temperature route to produce ceramics. Most metal alkoxides react with water to yield hydrous metal oxides which give the metal oxide on heating. Two or more metal alkoxides can be reacted to produce a mixed-metal oxide which yields binary or ternary oxides on hydrolysis.

The metal alkoxide route has been exploited largely to obtain oxide ceramic systems. Among nonoxide ceramics, the notable examples are polycarbosilane and aminosilane methods for SiC and Si_3N_4 production. Essentially, tetrachlorosilane ($SiCl_4$) when reacted with ethyl alcohol gives the alkoxide tetraethylortho silicate, also called tetraethoxysilane [$Si(OC_2H_5)_4$]. This alkoxide, acronym TEOS, serves as the starting material for making an inorganic polymer containing Si and O via the sol–gel route, i.e. silica glass or ceramic, or even glass-ceramic.

There are two ways of obtaining silica-based gels:

(i) destabilization of silica sol (pure or with metal ions added to aqueous solutions of salts) to obtain a homogeneous gel. When a sol is destabilized, the resultant product can be precipitates, unaggregated particles, or a homogeneous gel. It is the latter that we desire;

(ii) hydrolysis and polycondensation of metalorganic compounds (generally metal alkoxides) dissolved in alcohols and a small amount of water.

We give a brief description of these two methods:

(i) Destabilization of silica sol

A sol can be destabilized by increasing its temperature or by adding an electrolyte. A temperature increase reduces the quantity of intermicellar liquid by evaporation and increases the thermal agitation. This increases particle collisions

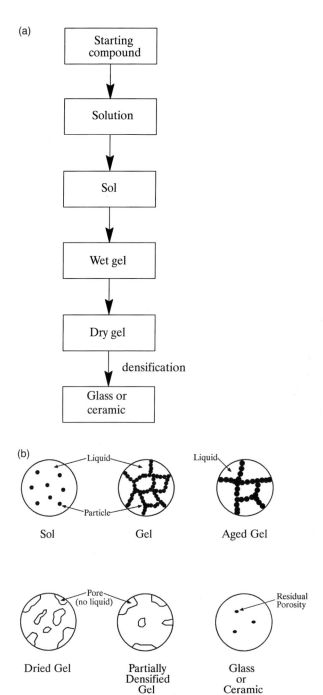

Figure 7.4   (a) Sol–gel process flow diagram. (b) Microstructural changes occurring during the sol–gel process.

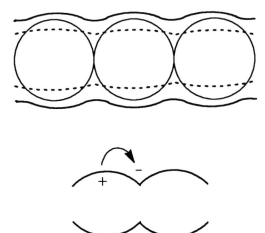

Figure 7.5 Strengthening of the particulate network.

and their linking by condensation of surface hydroxyls occurs. Electrolyte addition modifies the pH of the sol and reduces electric repulsion between particles. Addition of acid diminishes the pH to 5–6 and induces gel formation by *aggregation*. The sol is converted into gel progressively with microgel aggregates slowly invading the whole volume of sol. When about half of the silica has been converted to gel, a rapid increase in viscosity is noted.

An *aging* treatment results in a partial coalescence of particles and a strengthening of the network occurs. At the neck joining the particles there is a negative radius of curvature. Thus, local solubility at the neck is less than near the particle surface. Therefore, transport and deposition of silica occur preferentially to the neck region and *neck thickening* results. This results in a strengthening of the particulate network, Fig.7.5 (Zarzycki *et al.*, 1982).

Sol–gel transition occurs only if there are no active forces which promote coagulation into aggregates of higher silica concentration than the original sol. Metal cations, particularly polyvalent ones, tend to cause precipitation rather than gelling.

(ii) Hydrolysis and polycondensation of metalorganic compounds (metal alkoxides) (Zarzycki, 1984; Sakka, 1982).

This method involves three steps:

1. Mix the appropriate metal alkoxides (water-soluble salts if possible) and ethanol solutions to yield the desired cation ratio.
2. Hydrolyze the above mixture with a water based acid solution which can be diluted with alcohol. This polymerizes the mixture and a gel is formed in accordance with the following reaction

$$M(OR)_n + nH_2O \rightarrow M(OH)_n + nROH$$

where M is the metal and R is an alkyl group such as (C_2H_5). At this stage one can choose the conditions to obtain the bulk, powder, or fiber form of the precursor material.

3. Heat the gel slowly to obtain the oxide (glass or ceramic) through the reaction

$$M(OH)_n \rightarrow MO_{n/2} + \tfrac{1}{2}n\, H_2O$$

During the process of gel drying, the solvents (water, any residual organic material) are eliminated. Concomitantly, sintering of the porous oxide takes place. A well-controlled rate of drying is required because of the large shrinkage that accompanies the drying process. At an intermediate stage, a solid can be obtained with a very large amount of porosity on a microlevel. This is called xerogel. The glass or ceramic produced via the sol–gel process has virtually the same density, thermal expansion, refractive index, and mechanical characteristics as material of the same composition made by direct melting.

Fiber drawing

Fiber drawing is carried out in the course of sol to gel conversion at viscosities greater than 1 Pa s (1 Nm^{-2}s). The gelled fibers are heated to obtain glass fibers.

There are some very general but important points to be considered in gel spinning. A high molecular weight ($> 10^6$) is desirable. Spinning of dilute solutions into gels results in a minimum of polymer chain entanglement. Gelled fibers are highly porous and elastic. They should be stretched by drawing while in a temperature gradient. This removes the solvent and decreases the porosity.

When a tetraethoxysilane–water–hydrochloric acid–alcohol solution of appropriate composition is held at near room temperature, hydrolysis and polycondensation occur and the solution viscosity increases. Fibers can be drawn at a viscosity about 10 Pa s, which occurs during the course of sol to gel transition. It should be pointed out that for an alkoxide solution to be spinnable (Sakka, 1982) the solution must have linear polymers and an appropriate ratio of [H_2O]/[$Si(OC_2H_5)_4$], called the r-ratio. An r-ratio <2 gives a spinnable solution.

7.2.3   Optical glass fiber

The conventional method of making glass fiber described above is unsatisfactory for making optical grade fibers because of the extremely high purity required. Optical grade glass fiber is generally made by a vapor deposition process (Midwinter, 1979). Figure 7.6 shows two vapor deposition processes. High purity vapors of silicon and germanium are reacted to produce layers of glass, Fig. 7.6a. The composition of the glass can be changed by changing the proportions of Si and Ge, resulting in what is called a graded-index glass fiber. Silicon and ger-

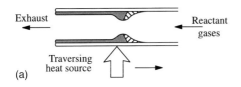

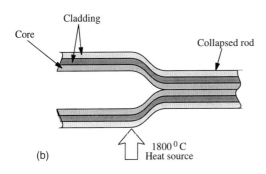

Figure 7.6 Two vapor deposition processes of making optical glass fiber. (a) High purity vapors of silicon and germanium are reacted to produce layers of glass. The composition of the glass can be changed by changing the proportions of Si and Ge to produce a graded-index glass fiber. (b) A modified chemical vapor deposition (MCVD) technique. The cladding is deposited on the inside of a hot silica tube. When sufficient cladding has been obtained, the reactants are changed to obtain the core glass. In the final stage, the temperature is increased and the tube is collapsed to form a solid preform rod.

manium are reacted inside a flame and the reaction products deposited on a rotating mandrel. Another variation involves reacting the components inside a glass tube. In either case, the preform so produced by the vapor deposition technique is heated and drawn into a fiber.

A modified chemical vapor deposition (MCVD) technique (Nagel *et al.*, 1982) is used to obtain a fiber consisting of a GeO_2–SiO_2 core and pure SiO_2 cladding, see Fig. 7.6b. In this case, first the cladding is deposited on the inside of a hot silica tube. When sufficient cladding has been obtained, the reactants are changed to obtain the core glass. In the final stage, the temperature is increased and the tube is collapsed to form a solid preform rod. This preform is converted into a fine filament by a drawing process and protective polymeric layers are applied.

Direct melting of optical glass fiber of an appropriate composition is also used under certain circumstances. Cladding and core rods are melted in a double crucible as shown in Fig. 7.7 and wound on a drum. A clean atmosphere must be maintained throughout all these operations to avoid the introduction of impurities. Fibers 5–10 km long (strain to fracture 2–3%) can generally be pulled. To protect the glass surface from handling–induced surface flaws, the pristine fiber is immediately coated with polymers. For applications involving communications, the coated fibers are gathered into cables for long haul and rough use such as in underground ducts, etc.

Optical glass fiber is a thin, flexible, and transparent guide through which light

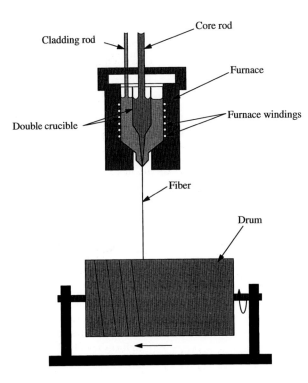

Figure 7.7   Direct melting process of making optical glass fiber. Cladding and core rods are melted in a double crucible and wound on a drum.

can be transmitted. Absorption and scattering of light traveling through such fibers result in signal attenuation. Intrinsic signal attenuation is a function of the wavelength and this component of optical loss is the lowest in GeO_2–SiO_2 glasses. Extrinsic absorption losses occur from transition metal and OH impurities. These are present to a considerable degree in the glass fiber produced by direct melting. In vapor phase deposited GeO_2–SiO_2 glasses, transition metal ion impurities can be reduced to <1ppb. OH absorption can be reduced by carefully preparing dry glass fibers. Complete elimination of OH is difficult.

Sources of strength loss in such fibers are mainly surface damage due to contamination and, possibly, the presence of microcracks (e.g. bubbles, etc.). Polymer surface coatings are used to minimize the damage. Fibers are also proof tested to breaking strains in the range of 0.5–1%.

Optical cable design

Optical fibers give a large bandwidth at small cross-section and should be under as little load as possible because of their brittleness. Essentially, we have an optical fiber core, a cladding, a buffer coating, and a sheath. The core fiber can vary in diameter from 5 to 1500 μm. The larger the core diameter, the more light the fiber can carry. Optical fiber fibers can be optimized to carry near IR, IR, UV or visible radiation. The light source can be traditional spectroscopic light or an intense high-power laser. The cladding, as described above, is a thin layer (5–10 μm) of glass or polymer with an index of refraction lower than that of the core so that it reflects the light back into the core. The buffer is also a polymer that protects the core and cladding from environmental hazards such as moisture, scratches and general handling. The sheath can be metallic or polymeric. Frequently, fiber reinforced polymers are used to allow for handling during cable manufacture as well as to allow for the design load. Polymeric composites such as aramid/epoxy or a hybrid of aramid and S glass in an epoxy matrix are frequently used. It gives added strength, stiffness and general protection from the environmental hazards.

Optical cable needs to be very rugged because of the abuse that it must take during a variety of cable forming techniques and rather harsh in-service environments (Gossing and Mahlke, 1987). It should be realized that when used for such purposes, optical fiber is exposed to aggressive environments that can cause loss of strength, time-dependent failure and/or premature failure. The mechanical reliability of optical fiber is thus very important. In particular, failure of glass fiber under the action of a constant stress, less than the initial strength, as a function of time can occur. Frequently, in the literature, this time-dependent failure is referred to as static fatigue.

Cable design must insure that the optical fibers in the cable maintain their optical properties (attenuation and dispersion) during service. The design must also minimize microbending effects, as explained in the following section. The term microbending refers to small-amplitude random bends of the axis of an optical fiber, with periodic components in the millimeter to centimeter range, which give rise to added transmission losses. It is also desirable to have a minimum differential thermal expansion or contraction during service in order to minimize microbending. In addition, the cable structure must be such that the fibers carry a load well below the proof test level at all times.

We mentioned above three types of optical fibers: step index single mode, step index multimode, and gradient index multimode. The last two types can carry more light that the step index single mode fiber. Lightness is a very important feature. Among the important features for such a system are the following (Clare, 1995):

- a convenient fiber optic system;
- flexible;
- small diameter;
- low loss;
- cheap and reliable.

7.3   Chemical composition

Silica-based glass fiber has been around for a long time. Common glass fiber is readily available commercially in a variety of different chemical compositions.

Most glass fibers are silica based (~50–60% SiO_2) and contain a host of other oxides of Ca, B, Na, Al, Fe, etc. They are commonly used for reinforcement of thermosetting and thermoplastic polymers. Table 7.1 gives the compositions of some commonly used glass fibers in composites. The designation E stands for electrical as E glass is a good electrical insulator besides having good strength and a reasonable Young's modulus; C stands for corrosion as C glass has a better resistance to chemical corrosion; S stands for higher silica content and S glass is able to withstand higher temperatures than others. It should be pointed out that more than 90% of all continuous glass fiber produced is of the E-glass type. Also included in Table 7.1 is the chemical composition of a special glass fiber called Cemfil.

The most important elements of an optical fiber are the core and the cladding. The refractive index of the core is higher than that of the cladding material such that total internal reflection occurs when the light hits the boundary with the low refractive index cladding. Optical glass fiber, in its simplest form, consists of a core of silica glass and a cladding of a refractive index lower than that of silica. Such an optical glass fiber is called a step-index fiber, because of the abrupt change in refractive index at the core/cladding boundary. A variety of optical fiber systems can be obtained. Figure 7.8 shows a schematic of optical guidance in what is called a multimode step-index fiber as well as the refraction index profile. There are three common types of optical glass fiber, multimode step index, multimode graded index, and single mode. The refractive indexes of these fiber types are shown in Fig. 7.9. The core fiber diameter can vary between 100 and 1500 μm in the case of a step-index fiber, between 5 and 600 μm for the graded-index fiber, and between 5 and 10 μm for the single-mode fiber.

A variety of sources such as bubbles, impurities, density changes, bends, etc. can result in attenuation of the signal in an optical glass fiber as shown in Fig. 7.10. There are, however, two important sources that limit the capacity of light transmission in any medium: absorption and scattering. In the case of glass

Table 7.1 *Chemical composition of some common glass fibers.*

Compound	E-glass	C-glass	S-glass	Cemfil
SiO_2	55.2	65.0	65.0	71.0
Al_2O_3	8.0	4.0	25.0	1.0
CaO	18.7	14.0	—	—
MgO	4.6	3.0	10.0	—
Na_2O	0.3	8.5	0.3	Na_2O+
K_2O	0.2	—	—	$K_2O=11$
Li_2O	7.3	5.0	—	<1.0

Refractive index

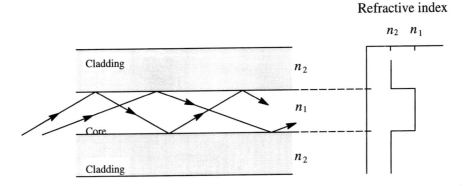

$n_1 > n_2 > n_{air}$

Figure 7.8 Schematic of optical guidance in a multimode step-index fiber and the corresponding refraction index profile.

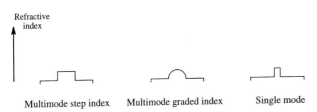

Figure 7.9 Refractive index profiles of three common types of optical glass fiber; multimode step index, multimode graded index, and single mode.

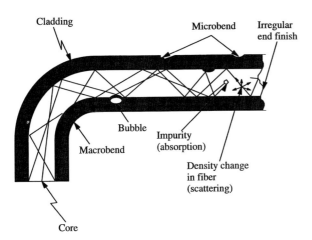

Figure 7.10 Some of the sources of signal attenuation in an optical glass fiber; bubbles, impurities, density changes, bends, etc.

fibers, absorption can be caused by a variety of impurities (ions of iron, copper, cobalt, vanadium, etc.) in very small amounts. This makes it imperative that the glass for optical fiber be extremely pure. Scattering refers to the change in light wave that is caused by density variations in the transmitting medium. Even tiny irregularities caused by temperature variation during cooling of the glass can cause an unacceptable amount of light scattering. Glass fiber for optical communication has a very controlled chemical composition. In a step-index fiber, both core and cladding are silica. Dopants such as GeO_2 and P_2O_5 are put in the silica for the core to increase the refractive index of the core. Dopants such as B_2O_3 and F can be used in the cladding to decrease the refractive index.

The core and cladding in an optical glass having different chemical compositions can result in an appreciable amount of thermal stresses. Consider, for example, a germanosilica optical glass fiber with the core and cladding having different coefficients of thermal expansion. When the glass is cooled below the fictive temperature, T_f, the core is put under high tensile stresses due to the thermal mismatch (Scherer, 1980).

7.4   Structure

Silica-based glass is a generic term representing an interesting and versatile class of materials. Glasses of various compositions can be obtained and they show very different properties. Structurally, however, all silica-based glasses have the same basic building block: a tetrahedron made up of four large oxygen ions with a silicon ion at the center of the tetrahedron. The composition of a single tetrahedron is thus SiO_4. Each oxygen ion is, however, shared by *two* tetrahedra, giving the bulk composition of SiO_2. It is this basic building block that is repeated in three dimensions in silica and silica-based materials. Different repeat patterns can result in different structures for the same composition. Thus, quartz is a crystalline form of pure silica while ordinary glass, which is amorphous, has a random network of silica tetrahedra. Figure 7.11 shows two different two-dimensional schemes of the network of silica tetrahedra: crystalline and amorphous. Glass has an amorphous structure, i.e. devoid of any long–range order so characteristic of a crystalline material. Silicon has a valency of four and bonds covalently to four neighboring atoms placed at the corners of a tetrahedron. In the silica tetrahedron, the silicon atom is at the center of the tetrahedron and oxygen atoms occupy the corners of the tetrahedron. Each of the oxygen atoms is shared by another silicon atom. Each polyhedron consists of oxygen atoms bonded covalently to silicon. The addition of other metal oxide types serves to alter the network structure and the bonding, and consequently the properties. The presence of crystalline regions is undesirable in a glass because such regions act as stress raisers. Glass fibers are quite isotropic; Young's modulus and thermal explosion coefficients are the same along the fiber axis and perpendicular to it. This, of course, is a result of the amorphous, three-dimensional network structure of glass, i.e. there is no preferential alignment of any structural units such as polymer chains along the fiber axis.

A great advantage of any silica-based glass is its ease of fabrication, which allows processes such as melt infiltration and compression molding to be used.

(a)

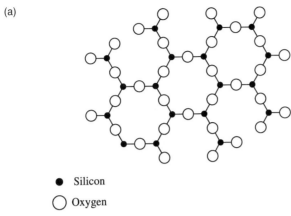

Figure 7.11 Two different two-dimensional network schemes of silica tetrahedra: (a) crystalline and (b) amorphous. Glass has an amorphous structure, i.e. devoid of any long-range order so characteristic of a crystalline material.

● Silicon

○ Oxygen

(b)

Glass has a low modulus, generally less than that of other common reinforcements. It also has a low failure strain.

7.5   Properties

Silica-based glasses generally have a very low density. The strength can be quite high, but the elastic modulus is not very high. Thus, while the strength-to-weight ratio of glass fibers is quite high, the modulus-to-weight ratio is only moderate. It would be safe to say that it is this low stiffness to weight ratio of glass that was the driving force for the development of the so-called advanced fibers, such as B, C, Al_2O_3, SiC, etc. Typical mechanical properties of different glass fibers are summarized in Table 7.2.

Although resistance to fire and many chemicals are very attractive features of glass, moisture can decrease the strength of glass fiber. In particular, the adsorption of moisture by freshly made glass fiber results in a drastic decrease in its strength. Glass fibers are also susceptible to what is called *static fatigue*, i.e. they cannot withstand loads for long periods of time.

Common glass fiber such as E-glass becomes severely corroded in an alkaline

atmosphere such as cement or concrete. Some special alkali-resistant glass fibers
have been developed. The properties of one such fiber, called *cemfil*, are given in
Table 7.2.

Carding (see Chapter 2) is a process involving mechanically opening,
combing, and aligning staple fibers to make nonwoven webs. The main advantage of this process is that nonwovens containing quite uniform fiber distribution
are obtained. The cardability of a staple fiber depends on its geometry, diameter,
length, frictional characteristics, crimp, etc. Crimp in a fiber is especially desirable. Common glass fiber requires the addition of a carrier polymeric fiber to be
carded. Without the polymeric fiber, conventional, crimpless glass fiber is too
rigid to be carded. A bicomponent glass fiber, called Miraflex glass fiber, has a
natural twist which allows it to be carded (Kenney *et al.*, 1995). Miraflex product
forms include yarn, staple fiber, fabric, mat and felt. It is soft to touch, feels silky,
is highly flexible and form-filling. Table 7.3 summarizes its characteristics.

Table 7.2 *Typical properties of some glass fibers.*

Glass fiber type	Density (g cm^{-3})	Tensile strength (GPa)	Young's modulus (GPa)
E	2.54	1.7–3.5	69–72
S	2.48	2.0–4.5	85
C	2.48	1.7–2.8	70
Cemfil	2.70	—	80

Table 7.3 *Some important characteristics of Miraflex fiber.[a]*

Diameter (μm)	7–8
Density (g cm^{-3})	2.46
Fiber length ratio (stretched/relaxed)	2
Specific heat (J/gK, at 22°C)	0.825
Specific heat (J/gK, at 200°C)	1.04
Softening point (°C)	665–690
Max. continuous use temperature (°C)	450
Thermal conductivity (W/mK)	1
Tensile strength/single filament (GPa)	0.7–1
Young's modulus (GPa)	67

Note:
[a] Data from Owens-Corning Co.

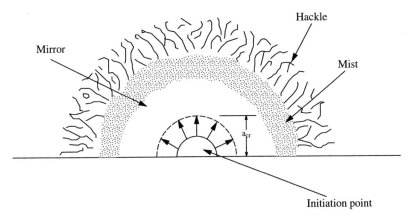

Figure 7.12   Schematic of the four regions observed on the surface of a glass fiber.

Fracture of glass fibers

Many brittle, amorphous materials show some tell tale markings on their fracture surfaces, e.g. many thermosetting polymers and silica-based inorganic glasses. In particular, Mecholsky *et al.* (1977, 1979) and Chandan *et al.* (1994) related fracture surface markings on optical glass fiber to the fiber strength and the time to failure. Figure 7.12 shows schematically the four distinct regions observed on the fracture surface of a glass fiber. These regions surrounding the point of fracture origin are indicated in Fig. 7.12 as a smooth mirror region, a mist region of small radial ridges, a hackle region consisting of large ridges, and finally a region of extensive crack branching. Extensive crack branching occurs beyond the hackle region. It turns out that the product of strength, σ, and the square root of the distance of each of these regions from the origin of fracture, is a constant. One can write

$$\sigma \, R_i^{0.5} = A_i$$

where R_i the radius of the mirror-mist, mist-hackle, or crack branching boundaries. These radii can be related to the initial flaw-depth, a, or the half-width, b, through the fracture mechanics analysis:

$$c/R_i = K_{\mathrm{Ic}} \, Y^2 / 2A_i^2$$

where $c = (ab)^{0.5}$, Y is a geometrical constant and K_{Ic} is the critical stress intensity factor or fracture toughness.

Lifetime prediction of glass fibers

It is of interest to be able to predict the life of a glass fiber, say an optical glass fiber in a cable buried under the sea. The starting point is crack growth velocity versus stress intensity factor, K. Generally, we can express crack velocity, v as

$$v = A K_I^n \qquad (7.4)$$

where K_I is the stress intensity factor, A is a constant, and n is the exponent. K_I is given by the following expression

$$K_I = Y\sigma\sqrt{a} \qquad (7.5)$$

where σ is the far field applied stress, a is the crack size, and Y is a geometric factor. From Eq. (7.4) we obtain an expression for the lifetime of a component that obeys this relationship. For example, a glass fiber component in an optical communication cable. Combining Eqs. (7.4) and (7.5), we can write for the time-to-failure, t_f

$$t_f = \int_{a_i}^{a_c} \frac{da}{v} = \frac{1}{A} \int_{a_i}^{a_c} \frac{da}{(Y\sigma\sqrt{a})^n} \qquad (7.6)$$

where a_i is the initial crack size and a_c is the critical crack size. Integrating Eq. (7.6), we obtain

$$t_f = B\sigma_c^{n-2}\left[1 - \left(\frac{\sigma}{\sigma_c}\right)^{n-2}\right] \approx B\sigma_c^{n-2}\sigma^{-n}$$

where

$$B = \frac{2}{A\,Y^2(n-2)K_{Io}^{n-2}}$$

and σ_c is the inert strength (strength in the absence of subcritical crack growth) and K_{Io} is the stress intensity factor at the onset of unstable crack extension.

The inert strength, σ_c, is given by a Weibull distribution (see Chapter 10) with the cumulative frequency

$$F(\sigma_c) = 1 - \exp[-(\sigma_c/\sigma_{co})^\beta]$$

where σ_{co} and β are the Weibull parameters. From Eq. (7.6), it follows that the lifetime (t_f) can be described by a Weibull distribution

$$F(t_f) = 1 - \exp[-(t_f/t_{fo})^\beta]$$

where $t_{fo} = B\,\sigma_{co}^{n-2}\,\sigma^{-n}$.

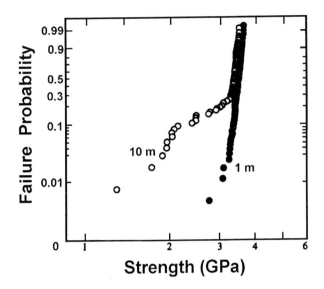

Figure 7.13 Failure probability of 1 m and 10 m gage length optical glass fibers. The long lengths of optical glass fibers have multiple flaw populations, i.e. there is more than one source of flaws, thus they do not follow the simple Weibull distribution (after Maurer, 1985).

It should be pointed out that Weibull statistics describe well the distribution of flaws in short lengths (a few centimeters long) of optical glass fibers but for long lengths (several meters or more), the strength distribution is not well described by a Weibull distribution. Figure 7.13 shows the failure probability of 1 m and 10 m gauge length optical glass fibers (Maurer, 1985). The reason for this is that long lengths of optical glass fibers have multiple flaw populations, i.e. there is more than one source of flaws.

7.6 Applications

Glass fiber is available in a variety of shapes and forms. Some of the important ones include continuous fiber and roving, staple fiber and chopped strand mat. Staple fibers are strands of individual filaments 200 to 400 mm long and are excellent for providing bulkiness for filling, filtration, etc. Chopped strand glass fibers consist of fibers chopped to various lengths, 3 mm to 50 mm, mainly for the purpose of mixing with a resin for making composites. Glass fiber mats consist of randomly dispersed chopped fibers or continuous fiber strands, held together with a resin.

Glass fibers find applications in a vast array of markets.

Automotive market. The automobile industry is one of the largest users of glass fiber. Polymer matrix composites containing glass fibers are used to make external body panels, bumper beams, pultruded body panels and airducts, engine components, etc. Parts made of glass fiber reinforced polymers are much lighter than metallic ones, making the automobile fuel efficient.

Aerospace market. Composites containing glass fibers have been used to make various aircraft parts. Examples include wings, helicopter rotor blades, engine ducts, etc. Glass fiber has a relatively low elastic modulus. Hence, it more common to use glass fiber reinforced polymer composites in the interior of an airplane rather than for primary structural parts. The radar transparency characteristic of glass has given it some key uses in the radar-evading 'stealth' technologies.

Marine market. Sail boats and boats, hulls and decks of commercial fishing boats and military minehunters are frequently made of glass fiber reinforced polymers. Glass fiber reinforced polyester is commonly used in making boats of all sizes.

Civil construction. This is another large market where glass fiber is widely used in one form or another. Typical applications include the use of glass fibers in polymeric resins for paneling, bathtubs and shower stalls, doors, windows etc. Glass fibers are also used as a reinforcement in a variety of household items such as paper, tapes, lamp shades, etc. Teflon coated glass fiber fabrics are used for domes and building covers of stadia, etc. Denver airport as well as Riyadh airport in Saudi Arabia have made extensive use of such fabrics. Some special alkali-resistant glass fibers have been developed for reinforcement of cement and concrete. Commonly, steel bars are used for such purposes. Cement, however, is very alkaline. An ordinary glass fiber such as E-glass will be severely corroded in an alkaline atmosphere, hence the need for special, alkali-resistant glass fiber (Majumdar, 1970; Hannant, 1978). The bonding between glass fiber and cement is mainly mechanical and has its origin in the shrinkage that occurs when the cement is set.

Insulation. Short glass fibers of random lengths, called glass wool, are used extensively for thermal and acoustical insulation as well as for filtration purposes. Air, trapped in the interstices of the pack, provides the insulation characteristics. Wall and ceiling panels in buildings and vehicles make use of such characteristics. Glass mat, consisting of shredded silica wool, needle punched with ordinary fiber glass to give structural integrity, is widely used for thermal insulation. Such a nonwoven mat is lightweight and flexible, it can also be conformed to different shapes for insulation purposes. Glass fiber can also be made into paper. Such a paper consists of silica fibers with about 2% binder to provide physical integrity. Fabrics made of high silica glass can be used as protective blankets in welding and burning processes. Such fabrics must be resistant to molten metals such as steels and should be usable in high temperature environments without producing smoke or noxious fumes. Woven, braided, or knitted fabrics up to 150 cm wide are available for such purposes.

High temperature textiles consisting of continuous filaments of amorphous silica in a variety of forms are available commercially. These can be used to replace asbestos-based high temperature textiles because of the health hazards associated with asbestos fiber. These high silica fibers are used for a variety of high temperature thermal insulation products: for example, welding curtains, furnace curtains, insulation for pipe and power plant cables, thermocouple insulation, fire rescue and smothering blankets. The Sandtex type silica fiber can be used up to temperatures as high as 1100°C. Knitted fabrics are more compliant and can be conformed to flat, curved, or other complex shapes. Knitted fabrics can be cut and sewn to many shapes unlike nonwovens. Ropes (twisted and braided) can also be made. A braided sleeve is easy to apply and fit. Heat insulation cordage can be braided onto a heating element.

Sporting goods. Not surprisingly, the supporting goods industry was one of the first to make use of glass fiber reinforced composites. Examples include bicycle frames, tennis rackets, golf shafts, cricket bats, skis, etc. Braided fibers in a resin matrix give high torsional stiffness to skis.

Electrical/electronic market. Glass fibers are used extensively in printed circuit boards, industrial circuit breakers, conduits for power cables, etc.

Optical communications. Communication systems that use lightwave transmission via fiberoptic cables are in use in many countries. They are more economical than conventional copper wire, radio, or satellite-based systems. Optical fibers carry much more information than copper wire of the same diameter. The reader will gain a good idea of the extent of the importance of glass fiber in the communication business if he/she is reminded that by the mid-1990s, more than 10^7 km of optical fiber had been installed in the US for long-distance communication. Optical fibers are also used in waveguides for long-distance transmission of digital signals generated by pulsed lasers or photo-diodes.

Sensors. Optical fiber sensors can be used to measure a variety of mechanical and physical parameters such as force, elongation, motion, velocity, rotation, temperature, current and voltage. Generally, such a sensor will consist of a light source (e.g. a light emitting diode or laser), an optical fiber to transmit the light to the transducer, a transducer, an optical fiber for the return transmission, and a detector. The transducer has the function of modifying one of the characteristics of light, e.g. intensity, wavelength, phase or polarization.

An unconventional application of optical glass fibers is for temperature measurement. The principle is very simple. A sensor is coupled to a flexible optical fiber which conducts light emitted by the target to a photo-detector. The photo-detector, a part of an electronics unit, converts the light signal to an electrical signal which is then digitized. The digitized electrical signal is converted to a temperature reading.

Optical fibers in medical field. Optical glass fibers have become very important in many medical fields where laser-assisted surgery is required. Laser surgery requires a beam directing capability, i.e. one needs a flexible delivery system to direct the energy for surgical purposes at a generally inaccessible place or site. Although generally glass fibers are used for such applications, single crystal and polycrystal ceramic fibers are used under special circumstances. Glass fiber is easy to fabricate, and one has great flexibility in chemical composition. Silica-based glass fibers, as described above, are the most common type used in laser surgery. This, of course, follows from its success in the telecommunications where the low loss is very important (0.1 dB/km).

An important requirement for laser surgery is a beam directing capability, which involves a flexible delivery system. A fiberoptic delivery system is a very convenient system for this purpose: it is flexible, low loss, small diameter, and allows deep, invasive laser endoscopy. One should add that there is always the potentially useful feature of using fibers as biosensors, i.e. one can have a coupled sensing and actuation device.

Chapter 8

Carbon fibers

Carbon fibers have become established engineering materials. In view of their commercial importance, we devote a separate chapter to them. Carbon is a very versatile element. It is very light, with a theoretical density of $2.27 \, \mathrm{g \, cm^{-3}}$. It can exist in a variety of forms, glassy or amorphous carbon, graphite, and diamond. Carbon in all these forms can be found in nature. Carbon in the graphitic form has a hexagonal structure and is highly anisotropic. The diamond form of carbon has a covalent structure and is an extremely hard material. The latest addition to the variety of forms of carbon is the Buckminster Fullerene, or the Buckyball with a molecular composition such as C_{60} or C_{70}. In this chapter, we follow the same sequence as in previous chapters; processing, structure, properties and applications of carbon fibers. However, in order to understand these aspects of carbon fiber, it is helpful to review the basic structure and properties of graphite.

8.1   Structure and properties of graphite

Carbon fiber is a generic name representing a family of fibers. Over the years, it has become one of the most important reinforcement fibers in many different types of composites, especially in polymer matrix composites. It is an unfortunate fact that the terms carbon and graphite are used interchangeably in commercial practice as well as in some scientific literature. Rigorously speaking, graphite fiber is a form of carbon fiber that is obtained when we heat the carbon fiber to a temperature greater than 2400°C. This process, called graphitization,

c

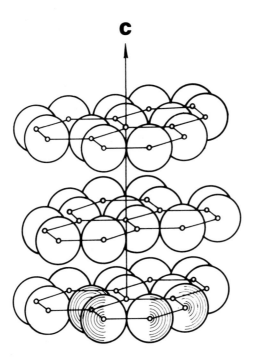

Figure 8.1   The atomic arrangement of carbon atoms in the hexagonal structure of graphite.

results in a highly oriented, layered crystallographic structure, which, in turn, leads to significantly different chemical and physical properties from non-graphitic forms of carbon (see Section 8.2). The atomic arrangement of carbon atoms in the hexagonal structure of graphite is shown in Fig 8.1. The *c*-axis is perpendicular to the basal plane. A graphite single crystal will thus have hexagonal symmetry and its elastic properties will be transversely isotropic in the layer plane. Such a crystal symmetry requires five independent elastic constants. Specifically, Young's modulus $E(\theta)$, as a function of angle θ, the angle between the *a*-axis and the stress axis, for a hexagonal crystal is given, in terms of the compliance, S_{ij}, by the following expression:

$$1/E(\theta) = S_1 \cos^4\theta + S_{33}\sin^4\theta + (S_{44} + 2S_{13})(\sin^2\theta\cos^2\theta) \qquad (8.1)$$

Table 8.1 gives the compliance (*S*) and stiffness (*C*) values of a graphite single crystal (Kelly, 1981). Note the large difference between C_{11} (very high) and C_{33} (very low) values as well as the low shear modulus, C_{44}. If we put the compliance values in Eq. (8.1), we can obtain a curve of the calculated tensile Young's modulus E as a function of θ (angular displacement from the *a*-axis) for a single crystal of graphite, Fig. 8.2. This figure shows clearly the highly anisotropic nature of a graphite single crystal. Clearly, the more aligned the basal planes in a carbon fiber, i.e. the more graphitic the structure, the higher will be the modulus in the direction of the fiber axis. We shall see presently that a nearly perfect align-

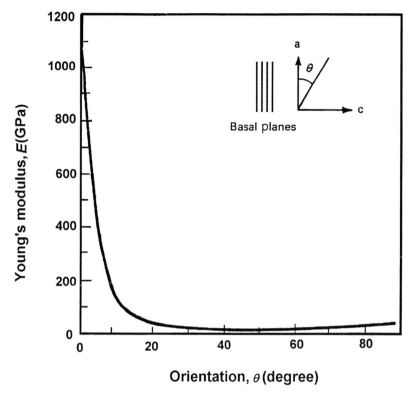

Figure 8.2 Calculated tensile Young's modulus E as a function of θ (angular displacement from the a-axis) for a single crystal of graphite.

ment of graphitic basal planes can be obtained in mesophase-pitch-based carbon fibers, which in turn results in a Young's modulus in the fiber direction as high as 85% of the theoretical maximum. In contrast, a nonmesophase-pitch-based carbon fiber having poor alignment of basal planes can have a Young's modulus as low as 5% of the theoretical value (Edie and Dunham, 1989; Edie, 1990)

Table 8.1 *Compliance (S) and stiffness (C) values for a graphite single crystal (after Kelly, 1981).*

S (GPa^{-1})	C (GPa)
$S_{11}=0.00098$	$C_{11}=1060$
$S_{33}=0.0275$	$C_{33}=36.5$
$S_{44}=0.25$	$C_{44}=4$
$S_{12}=-0.00016$	$C_{12}=180$
$S_{13}=-0.00033$	$C_{13}=15$

8.2   Processing of carbon fibers

Carbon fiber is fabricated by controlled pyrolysis of an organic fiber precursor. Some of the commercially important precursors, their chemical structure, and the carbon fiber yield are given in Table 8.2 (Riggs, 1985). Depending on the precursor and processing, one can obtain a variety of carbon fibers with different strength and modulus. Important types are high strength (HS), high modulus (HM), intermediate modulus (IM) and super high modulus (SHM).

Polyacrylonitrile (PAN) precursor fibers are more expensive than rayon. Nevertheless, PAN is more commonly used because the carbon fiber yield is about double that from rayon. Pitch-based carbon fibers are also important, because, potentially pitch is perhaps the cheapest raw material. Table 8.2 shows that carbon yield is highest from the mesophase pitch. The reader is cautioned that this is true only if we exclude the losses during the mesophase conversion step. If, however, one compares the overall carbon fiber yield from raw pitch to that from PAN, then the yield from PAN is higher. In any event, the carbon fiber yield or precursor weight loss is a very important factor in the economics of processing.

There are certain essential steps common to all processes of carbon fiber manufacture. These are:

(a) *Fiberization.* This involves converting the precursor material into a fibrous form, i.e. extrusion of a polymer melt or solution into a precursor fiber.

(b) *Stabilization.* This treatment renders the precursor infusible during the subsequent high temperature processing. This involves preoxidation or cross–linking of the polymer by thermosetting and is done at relatively low temperatures (200–450°C), usually in air.

(c) *Carbonization.* This step involves conversion of the stabilized precursor

Table 8.2   *Some commercially important precursors, their chemical structure, and the carbon fiber yield (after Riggs, 1985).*

Precursor	Structure	Yield (wt%)
Rayon	$(C_6H_{10}O_5)_n$	20–25
PAN	$(CH_2-CH)_n$	45–50
Mesophase pitch	CN	75–85[a]

Note:
[a] Excluding the losses in the initial mesophase conversion step before precursor fiber preparation.

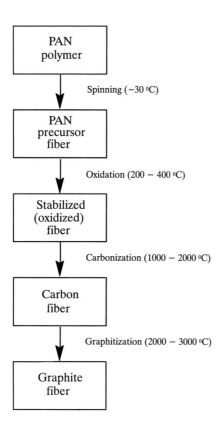

Figure 8.3 Flow diagram of the fabrication of PAN-based carbon fiber.

fiber into a carbon fiber. This is carried out in an inert atmosphere (pure N_2 generally) at a temperature between 1000 and 2000°C. At the end of this step the fiber has 85–95% carbon content.

(d) *Graphitization.* This is an optional treatment. It is done in Ar or N_2 at a temperature between 2400 and 3000°C. This step increases the carbon content to more than 99% and imparts a very high degree of preferred orientation to the fiber.

8.2.1   PAN-based carbon fiber

Polyacrylonitrile (PAN) is the most common precursor used to make carbon fibers. A flow diagram showing the steps involved in making PAN-based carbon fiber is shown in Fig. 8.3. The PAN precursor has a flexible polymer chain structure like any other polymer, but it has an all carbon backbone chain that contains polar nitrile groups, as shown in Fig. 8.4. During the stabilization treatment, the PAN precursor fiber is heated to 200–220°C, under tension. When this is done oxygen is absorbed, and it serves to cross-link the chains; the fibers turn black, and a stable ladder structure is formed. A ladder polymer is a rigid

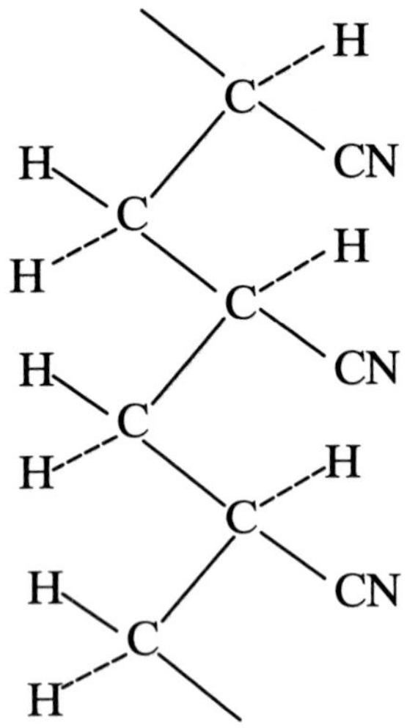

Figure 8.4 The flexible polymer chain structure of PAN precursor. Note the all-carbon backbone chain.

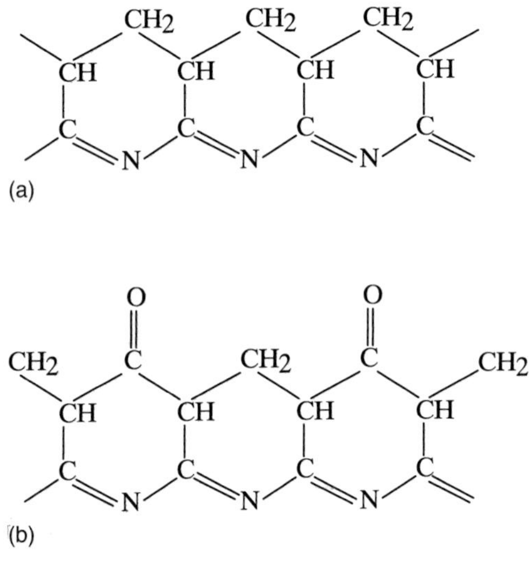

(a)

(b)

Figure 8.5 Structural changes occurring in PAN during stabilization. (a) The ladder structure obtained when PAN is heated in the absence of oxygen; (b) the structure obtained when it is heated in the presence of oxygen. Note the oxygen fixed to the ladder structure in (b).

and thermally stable structure. This treatment, done under tension, helps in maintaining the orientation of the ring structure during subsequent processing. Figure 8.5a shows the ladder structure obtained when PAN is heated in the absence of oxygen, while Fig. 8.5b shows the structure obtained when it is heated in the presence of oxygen. Note the oxygen fixed to the ladder structure. During the next stage of carbonization, the stabilized precursor fiber is heated between 1000 and 1500°C, which results in the formation of well-developed hexagonal networks of carbon. A considerable quantity of gases is evolved. This gas evolution is partly responsible for some crack formation in the carbon fiber and, consequently, lower tensile strength. Table 8.3 presents the property data for four different types of commercial PAN-based carbon fibers.

Commercially produced carbon fibers, invariably, have a *size*, i.e. a protective surface coating. A size is applied to provide ease of handling and improved adhesion between carbon fibers and a polymeric matrix material. The uncoated carbon fiber picks up surface charge easily when it comes in contact with rubbing surfaces such as rollers, pulleys, guides, spools, etc. Handling such fibers (winding, weaving, or braiding) can cause breakages, and these tiny fragments can become airborne and short-circuit electrical machinery. The major market for carbon fibers is with PMCs. It is, therefore, understandable that a low

Table 8.3 *Properties of three different types of commercial PAN-based carbon fibers.*[a]

Manufacturer	Trade name	Young's modulus, E (GPA)	Tensile strength, σ (GPa)	Strain to failure, ϵ (%)
High modulus fiber				
Celanese	GY-70	517	1.86	0.4
Hercules	HM-S	350	2.21	0.6
Hysol Grafil	HM	370	2.75	0.7
Toray	M 50	500	2.50	0.5
Toray	M 55J	540	3.63	0.7
Intermediate modulus fiber				
Celanese	Celion 1000	234	3.24	1.4
Hercules	IM-6	276	4.40	1.4
Hysol Grafil	Apollo IM 43-600	300	4.00	1.3
Toho Beslon	Sta-grade Besfight	240	3.73	1.6
Union Carbide	Thornel 300	230	3.10	1.3
Toray	M 30	294	3.92	1.3
High strength fiber				
Celanese	Celion ST	235	4.34	1.8
Hercules	AS-6	241	4.14	1.7
Hysol Grafril	Apollo HS 38-750	260	5.00	1.9
Toray	T 800	300	5.70	1.9
Toray	T 1000	294	7.06	2.4

Note:
[a] All the numbers in these tables are from manufacturer's data.

molecular weight epoxy-based coating (without the hardener) is commonly used as a size. An oxidative surface treatment is also given to improve adhesion. This produces a rough etched surface having reactive oxide sides. These aid in mechanical and chemical bonding with a polymeric matrix. Before using carbon fibers in a metal or ceramic matrix, any size on carbon fibers must be removed by burning off.

8.2.2 Pitch-based carbon fibers

Pitches form an important and low cost raw material for producing carbon fiber. There are three common sources of pitch:

- petroleum asphalt;
- coal tar;
- polyvinyl chloride (PVC).

Pitches are thermoplastic in nature and are difficult to carbonize without being first stabilized against melting during pyrolysis.

A schematic of the process of making carbon fibers from a pitch is shown in Fig. 8.6. It involves the following steps:

- extrude or melt spin pitch into a fibrous form;
- stabilize at between 250°C and 400°C;
- carbonize;
- graphitize.

Spinnability of the pitch and its conversion into a nonfusible state are the most important steps. These properties depend on the chemical composition and molecular weight distribution of the pitch. The pitch composition depends on its source. In fact, there is a tremendous variability in pitch because it is a mixture of hundreds of different species, varying with the crude source and the process conditions in the refinery. Indeed, this is one of the main problems in the manufacture of pitch-based carbon fibers. Generally, pitches are made up of different organic compounds containing groups of condensed benzene-ring systems separated by and carrying alkyl chains. Petroleum and coal-tar pitches may also contain large amounts of sulfur. PVC pitch is also a mixture of several different organic compounds—polynuclear aromatic compounds. Polynuclear aromatic compounds have three or four aromatic nuclei, i.e. consisting of three or four rings per molecule. One suggested formula is $C_{62}H_{52}$. PVC pitch is obtained by thermal degradation of PVC at 400°C under nitrogen for 30 minutes, the yield being 20%. The melting range is 150–200°C, and the molecular weight is between 700 and 800.

The suitability of pitch for conversion to carbon fiber depends on a number

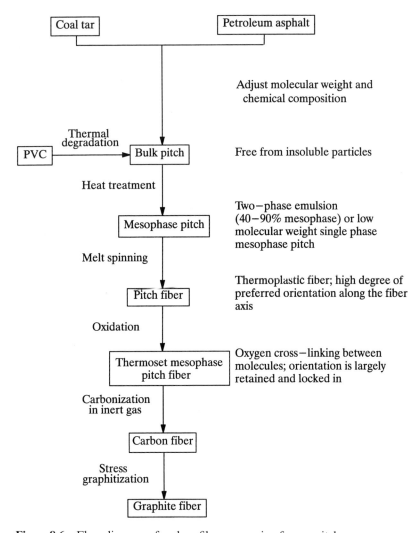

Figure 8.6   Flow diagram of carbon fiber processing from a pitch.

of factors. According to Singer (1979), the pitch should have a high carbon content (>90%), high aromatic content (>50%), low impurities, and molecular weight and molecular weight distribution, viscosity and rheological character-istics consistent with ease of spinning, followed by easy conversion to an infusible state.

Spinning and rheology of mesophase pitch

Although melt spinning is commercially popular, centrifugal spinning and jet spinning can also be used. Mesophase pitch, a thermoplastic, has a nematic liquid crystal structure, i.e. its molecules are rigid rodlike. The two-phase pre-cursor pitch is agitated prior to spinning to form a homogeneous mixture and spun into filaments in the temperature range where the viscosity varies from 1 to 20 Pa s (Nm^{-2}s). Fibers can be spun at speeds of 3–100 m min^{-1} with diameters of 10–20 μm and having the same composition as the mesophase pitch. The as-spun mesophase fibers are anisotropic because of the nematic liquid crystal structure. These fibers have large, elongated, anisotropic domains (about 4 μm diameter) aligned along the fiber axis and are thermoplastic in nature. The anisotropic domains can easily be viewed under polarized light. Fibers drawn from isotropic pitches or those drawn from rayon or acrylic (PAN) precursor do not show such anisotropic domains. Oxidation treatment is given to stabilize thermally against internal relaxation and to render the fibers infusible. The rest of the process is essentially similar to PAN-based carbon fiber fabrication, except that stress is applied during graphitization.

8.3   Structure of carbon fibers

As we have said above, a very characteristic feature of the structure of carbon fiber is the alignment of the graphitic planes along the fiber axis. The degree of alignment of these graphitic planes can vary depending on the precursor used and the processing, especially the heat treatment temperature used. When viewed in a transmission electron microscope, the heterogeneous microstructure of carbon fibers becomes clear. In particular, the pronounced irregularity in the packing of graphitic lamellae from the fiber surface inward can be seen. The basal planes are much better aligned in the near surface region of the fiber. This region is called the *sheath.* The material inside the sheath can have a radial structure or an irregular layer structure, sometimes termed the *onion skin* structure (Inal *et al.*, 1980). The radial core and well-aligned sheath structure is more commonly observed in mesophase-pitch-based carbon fibers. A variety of arrangement of graphitic layers can be seen. Figure 8.7 shows schematically some of these (Herakovich, 1989). In very general terms, the graphitic ribbons are oriented more or less parallel to the fiber axis with random interlinking of layers, longitudinally and laterally (Jain and Abhiraman, 1987; Johnson, 1987; Deurbergeu and Oberlin, 1991). Figure 8.8a shows a two-dimensional representation of this lamellar structure (*turbostratic* structure), while Fig. 8.8b shows lattice fringes in a high resolution transmission electron micrograph of a high modulus PAN-based carbon fiber (Johnson, 1987). A schematic of the three-dimensional structure of PAN-based carbon fiber is shown in Fig. 8.8c (Bennett *et al.*, 1983). Note the distorted carbon layers and the irregular space filling. The degree of alignment of the basal planes increases with the final heat treatment temperature. Examination of lattice images of the cross-section of carbon fiber shows essentially parallel basal planes in the skin region, but extensive folding of layer planes can be seen in the core region. It is thought that extensive inter-

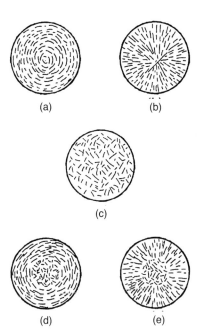

Figure 8.7   Some ideal-
ized cross-sections of
carbon fibers (after
Herakovich, 1989): (a) cir-
cumferentially
orthotropic; (b) radially
orthotropic; (c) trans-
versely isotropic; (d) cir-
cumferentially orthotropic
with transversely isotropic
core; (e) radially
orthotropic with trans-
versely isotropic core.

linking of lattice planes in the longitudinal direction is responsible for the better
compressive properties of carbon fiber than aramid fibers (see Chapter 4).

In spite of the better alignment of basal planes in the skin region, the surface
of carbon fibers can show extremely fine-scale roughness. A scanning electron
micrograph of AS4 carbon fibers is shown in Fig. 8.9a, while an atom-force
microscope picture of the same fibers is shown in Fig. 8.9b. Note the surface
striations and the roughness at a microscopic scale.

In general, mesophase-pitch-based carbon fibers have more aligned crystallites
than PAN-based carbon fiber. Robinson and Edie (1996) studied the effect of
mesophase pitch precursor and capillary entry design on the structure and thermal
characteristics of ribbon-shaped fibers. They observed that carbon fiber derived
from a naphthalene-based mesophase showed a higher degree of crystallinity, preferred orientation, and graphitization, and a lower electrical resistivity than a
carbon fiber obtained from a petroleum-based mesophase fiber. An important
finding of these authors was that the spinneret capillary entry design was very
important to the development of a linear transverse structure. It was this linear-
ized transverse structure that resulted in very low electrical resistivity.

8.4   Properties of carbon fibers

As described above, a carbon fiber with a perfectly graphitic structure will have
a theoretical Young's modulus of slightly over 1000 GPa. In practice, however,

(a)

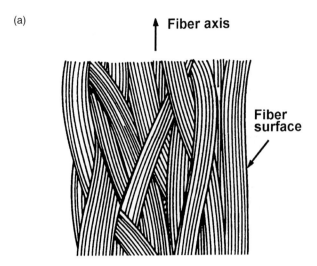

Fiber axis

Fiber surface

Figure 8.8 (a) Two-dimensional representation of the lamellar structure of carbon fiber; (b) lamellar structure of carbon fiber as seen in TEM (courtesy of D.J. Johnson).

(b)

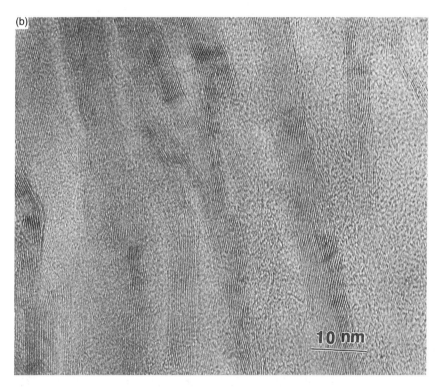

10 nm

Young's modulus is about 50% of the theoretical value in the case of PAN-based carbon fiber and may reach as much as 80% of the theoretical value for mesophase-pitch-based carbon fiber. The strength of carbon fiber falls way short of the theoretical value of 180 GPa (Reynolds, 1981). Practical strength values of carbon fiber may range from 3–20 GPa. The main reason for this is that while the modulus is determined mainly by the graphitic crystal structure, the strength

(c)

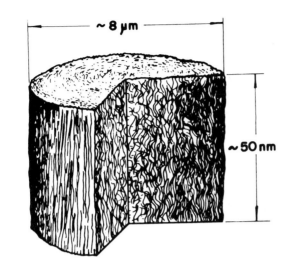

Figure 8.8 (c) three-dimensional structure of PAN-based carbon fiber (after Bennett *et al.*, 1983).

is a very sensitive function of any defects that might be present, for example, voids, impurities, inclusions, etc. The strength of carbon fiber thus depends on the gage length, decreasing with increasing gage length. This is because the probability of finding a defect in the carbon fiber increases with its gage length. Understandably, it also depends on the purity of the precursor polymer and the spinning conditions. A filtered polymer dope and a clean spinning atmosphere will result in a higher strength carbon fiber for a given gage length.

Following Hüttinger (1990), we can correlate the modulus and strength of carbon fiber to its diameter. We make use of Weibull statistics to describe the mechanical properties of brittle materials (see Chapter 10). Brittle materials show a *size effect*, i.e. the experimental strength decreases with increasing sample size. This is demonstrated in Fig.8.10 which shows a log–log plot of Young's modulus as a function of carbon fiber diameter for three different commercially available carbon fibers. The curves in Fig. 8.10 are based on the following expression:

$$(E/E_0) = (d_0/d)^n \tag{8.2}$$

where E is Young's modulus of the commercial carbon fiber of diameter d while E_0 is the theoretical Young's modulus and d_0 is the fiber diameter corresponding to E_0. The slope of the straight lines in Fig. 8.10 provides the exponent n. The value of n is about 1.5 and is independent of the fiber type. It would appear that these fibers attain their theoretical value of modulus at a diameter of about 3 μm. We can perform a similar treatment to the tensile strength property of carbon fibers. The corresponding equation will be

$$(\sigma/\sigma_0) = (d_0/d)^n \tag{8.3}$$

Figure 8.9 (a) Scanning electron micrograph of AS4 carbon fibers. The fiber diameter is 7μm; (b) atom-force microscope picture showing the fiber surface roughness at an extremely fine scale (courtesy of R.K. Eby).

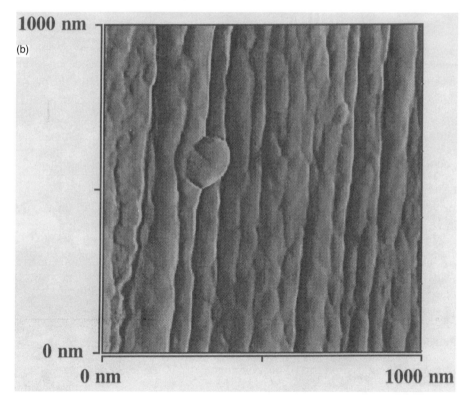

Now the theoretical strength of a crystalline solid, σ_0 is expected to be about $0.1E_0$ (Meyers and Chawla, 1984), i.e. in this case $\sigma_0 = 100$ GPa. For this value of σ_0, the exponent n in Eq. (8.3) turns out to be 1.65 and 2 (Hüttinger, 1990). This means that in order to obtain a strength of 100 GPa, the diameter of the carbon fiber must be reduced from d to $d_0 \approx 1$ μm. Note that this d_0 value corresponding to theoretical strength is less than the d_0 value corresponding to the

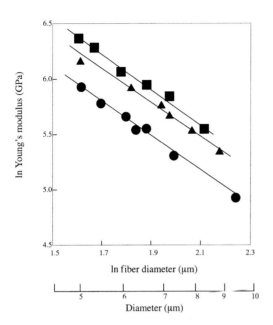

Figure 8.10 Log–log plot of the Young's modulus as a function of carbon fiber diameter for three different commercially available carbon fibers (after Hüttinger, 1990).

theoretical modulus. The strength corresponding to a 3 μm diameter carbon fiber from Eq. (8.3) will be between 12 and 18 GPa, an extremely high value. This can be understood in terms of the heterogeneous structure of carbon fiber. Recall from our discussion above that the near surface region of a carbon fiber has more oriented basal planes than the core. As we make the fiber diameter smaller, essentially we are reducing the proportion of the core to the near surface region.

Table 8.4 gives the properties of mesophase-pitch-based carbon fibers. The reader should note the high density and high modulus of pitch-based fibers compared to PAN-based fibers. The very high temperature treatment of graphitization (to a temperature as high as 3000°C) increases the degree of order in carbon fibers. This is accompanied by a large increase in the elastic modulus of the fiber. The tensile strength of PAN-based fibers when subjected to high temperature treatment, however, falls, Fig. 8.11 (Watt, 1970). This is attributed to the presence of discrete flaws on the fiber surface and within it. Most of the volume defects in carbon fibers originate from the following sources:

   (i) inorganic inclusions;
   (ii) organic inclusions;
  (iii) irregular voids from rapid coagulation;
  (iv) cylindrical voids precipitated by dissolved gases.

These defects are transformed during the high temperature treatment into diverse imperfections. Basal-plane cracks called *Mrozowski* cracks are perhaps

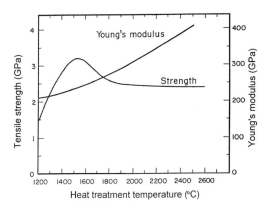

Figure 8.11 Tensile strength and Young's modulus of PAN-based fibers as a function of high temperature treatment (after Watt, 1970).

the most important flaw type that limits the tensile strength of carbon fibers. These occur as a result of anisotropic thermal contractions within the ribbon structure on cooling down from high temperature treatment (>1500°C). These cracks are generally aligned along the fiber axis. Their presence lowers the tensile strength of the fiber by providing easy crack nucleation sites. The fiber elastic modulus, however, is unaffected because the elastic strains involved in the modulus measurement are too small. Surface flaws can also limit the tensile strength of the carbonized fibers. Oxidation treatments tend to remove the surface defects and thus increase the strength levels of the fiber.

Figure 8.12 shows a comparison of tensile strength and Young's modulus for a variety of carbon fibers. It should be mentioned that the compressive strength

Table 8.4 *Tensile properties of mesophase pitch-based carbon fibers.[a]*

Manufacturer	Trade name	Young's modulus, E (GPA)	Tensile strength, σ (GPa)	Strain to failure, ϵ (%)
Union Carbide	Thornel P-25	140	1.40	1.0
	P-55	380	2.1	0.5
	P-75	520	2.1	0.4
	P-100	700	2.1	0.3
	P-120	820	2.20	0.2
Osaka Gas	Donacarbo	140	1.80	1.3
	F-140			
	F-600	600	3.00	0.5

Note:
[a] All the numbers are from manufacturer's data.

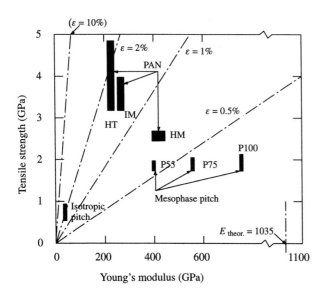

Figure 8.12 Comparison of tensile strength and Young's modulus for a variety of carbon fibers (after Hüttinger, 1990).

of carbon fiber is low compared with its tensile strength. The ratio of compressive strength to tensile strength for carbon fibers may vary anywhere between 0.2 and 1 (Kumar, 1989). High modulus PAN-based carbon fibers buckle on compression, forming kink bands at the inner surface of the fiber. A crack initiates on the tensile side and propagates across the fiber (Johnson, 1990). In contrast, high modulus mesophase-pitch-based carbon fiber deform by a shear mechanism leading to kink band formation at 45° to the fiber axis.

8.5 High thermal conductivity carbon fibers

Certain types of carbon fibers can have extremely high thermal conductivity. Examples of such carbon fibers include:

(i) A short carbon fiber produced by catalytic chemical vapor deposition (CCVD) (Baker and Harris, 1978). It is also called a vapor grown carbon fiber (VGCF).

(ii) A pitch-based, highly aligned, continuous carbon fiber.

We provide a brief description of these.

Vapor grown carbon fiber

This fiber is made by the decomposition of a suitable hydrocarbon (e.g. natural gas, acetylene or benzene) on a heated substrate in the presence of a catalyst. The catalyst is a transition metal such as iron, cobalt or nickel and the substrate can be a material such as carbon, silicon or quartz that can withstand a

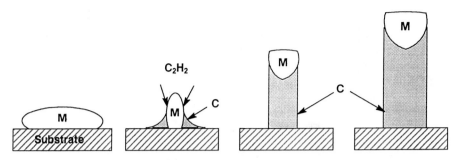

Figure 8.13   Formation of vapor grown carbon fibers via catalyzed decomposition of acetylene.

temperature greater than 1000°C. Short filaments, up to several hundred millimeters long and tens of micrometers in diameter are produced at temperatures between 300 and 2500°C. Figure 8.13 shows, schematically, the catalyzed decomposition of acetylene to produce vapor grown carbon fibers. A summary of the process of vapor grown carbon fiber formation is as follows (see Oberlin *et al.*, 1976; Tibbetts *et al.*, 1986; Tibbetts, 1989). The particles of the transition metal catalyst are reduced by hydrogen or their surface is cleaned such that polymerization and condensation of hydrocarbons occur, developing a hexagonal planar network of carbon. This planar network grows normal to the substrate surface in the space between the particle and the substrate. The particle is driven upward and the fiber grows. As the particle moves up, a hollow tube is left behind. Vapor deposition of carbon occurs on this hollow central filament which in turn grows by the catalytic action on the small transition metal particle. Carbon atoms diffuse through the catalyst and precipitate on the growing filament. The driving force for carbon diffusion is the gradient in the carbon concentration between the adsorbing surface of the catalyst and its precipitating surface. The filaments are thickened by vapor deposition of carbon on the external surface. An important point is that the carbon deposited in this manner results in a very high degree of orientation parallel to the plane. Thus, a vapor grown carbon fiber has the graphitic basal planes in a nested, rolled up form. A high temperature heat treatment at 3000°C and above can produce an extremely ordered internal structure which, at times, is reflected in some external faceting. Because of their dimensions and processing, some people have likened vapor grown carbon fibers to carbon whiskers. However, that is incorrect inasmuch as a whisker must be a single crystal and these short carbon fibers are not.

When grown at low temperatures, the VGCF takes a twisted or helical shape and the structure shows a pleated form, as shown in the SEM micrograph in Fig. 8.14. At high temperatures, relatively straight filaments are grown. Vapor grown carbon fibers are characterized by a high tensile strength (7 GPa), high modulus

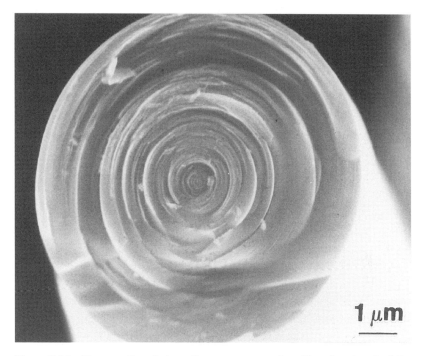

Figure 8.14   Cross-sectional view of vapor grown carbon fiber showing the foli-
ated structure (courtesy of G. Tibbetts).

(400 GPa), high density (2 g cm^{-3}) and exceptionally high thermal conductivity
(about four times that of copper) (Piraux *et al.*, 1984). Not unexpectedly, the
strength and modulus of these fibers decrease with increasing diameter of the
fiber.

Pitch-based high thermal conductivity carbon fiber

Amoco has developed a family of ultra high modulus continuous graphite fibers
and preforms with axial thermal conductivity to 1100 W/mK. The extremely
high thermal conductivity is a direct result of an extremely high degree of
crystallinity during carbonization of the mesophase pitch precursor fiber. Table
8.5 compares the thermal conductivity of some metals, a variety of carbon fibers
and some carbon fiber reinforced unidirectional composites. Also included are
specific thermal conductivity values. Note the extremely high specific thermal
conductivity values of K1100X fiber and their composites. These fibers find
applications in the following areas:

- high thermal conductive core materials for aircraft engines;
- heatsinks for military electronics;
- spacecraft structures requiring high stiffness and high thermal conductivity.

8.6   Hollow carbon fibers

There is some interest in obtaining a hollow carbon fiber which is expected to lead to improvements in toughness in carbon fiber reinforced polymer composites (Rhee, 1990). They may also find some applications as high temperature filters and in biomedical fields because of the highly selective absorption characteristics of the carbon surfaces. The starting material is anisotropic mesophase pitch. Two specially designed spinnerets have been used: (i) a spinneret with a C-shaped orifice; (ii) a spinneret with a tiny jet nozzle located at the center of the orifice. The wall thickness of the carbon fiber can be varied by controlling the draw rate and the pickup or winding speed. Generally, tensile strength increases with decreasing wall thickness of the carbon fiber.

8.7   Hazards of carbon fibers

Tiny fragments of carbon fiber, if released into the atmosphere, can pose a health hazard to human beings and can cause problems for electrical and electronic equipment. Some of these human health hazards are, as discussed in Chapter 6 with regard to asbestos fiber, common to all very small diameter fibers. They can cause skin and eye irritation and a variety of lung diseases. It turns out that in

Table 8.5 *Comparison of thermal conductivity of some metals, carbon fibers and their composites.*[a]

Material	Longitudinal thermal conductivity (W/mK)	CTE (10^{-6}/K)	Density (g cm^{-3})	Specific thermal conductivity (W/mK/g cm^{-3})
Al 6063	218	23	2.7	81
Copper	400	17	8.9	45
P-100	520	−1.6	2.2	236
P-120	640	−1.6	2.1	305
K1100X	1100	−1.6	2.2	500
K1100X/Al(55 v/o)	634	0.5	2.5	236
K1100X/epoxy(60 v/o)	540	−1.4	1.8	344
K1100X/Cu (46 v/o)	709	1.1	5.9	117
K1100X/C (53 v/o)	696	−1.0	1.8	387

Note:
[a] Data from Amoco Performance Products, Inc.

so far as the carcinogenesis potential is concerned, fiber shape, size, and aspect ratio are more important factors than chemical composition, although according to some NASA studies (NASA, 1978, 1979, 1980) it is hard to quantify the effects because of the paucity of data. Certain precautions are recommended during handling and processing of such fine fibers (Kowalska, 1982). Possible human exposure to such fine fibers can occur during fabrication, cutting, grinding, etc. Carbon fibers are also good electrical conductors, and if any released fibers settle on electrical or electronic equipment, they can cause short-circuiting.

8.8 Applications of carbon fibers

Carbon fibers are routinely used to reinforce low modulus polymeric materials. Applications of such composites range from aerospace to sporting goods. Carbon fiber is a highly inert material. This makes it difficult to have strong adhesion between carbon fiber and a polymer matrix. One solution is to make the fiber surface rough by oxidation or etching in an acid, giving a mechanical keying effect at the fiber/matrix interface. Many different grades of carbon fiber are produced by a large number of manufacturers worldwide. Because of its small diameter (5–7 μm), one can use traditional textile processing techniques (e.g. knitting, weaving and braiding) to produce a variety of shapes and forms, fine or heavy fabric, balanced or unbalanced construction, stiff or easy to drape and conforming to complex shapes!

An unusual application involves the use of carbon fiber to reinforce cement. This results in improved tensile and flexural strength, high impact strength, improved dimensional stability, etc.

Carbon fiber reinforced ceramic composites also find some important applications. Carbon is an excellent high temperature material when used in an inert or nonoxidizing atmosphere. In carbon fiber reinforced ceramics, the matrix may be carbon or some other glass or ceramic. Unlike other nonoxide ceramics, carbon powder is nonsinterable. Thus, the carbon matrix is generally obtained from pitch or phenolic resins. Heat treatment decomposes the pitch or phenolic to carbon. Many pores are formed during this conversion from a hydrocarbon to carbon. Thus, a dense and strong pore-free carbon/carbon composite is not easy to fabricate.

Mesophase-pitch-based carbon fibers can have up to three times the thermal conductivity of copper. This would make them an ideal material for thermal management applications, e.g. brake disks where heat dissipation is of prime consideration. The extremely high thermal conductivity is a direct result of the extremely high degree of crystallinity obtained during carbonization of the mesophase-pitch precursor fiber.

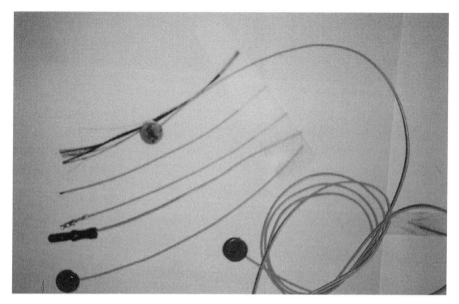

Figure 8.15   Electrode leads made of flexible, radiotranslucent carbon fiber (courtesy of Minnesotta Wire and Cable Co.).

Rayon-based carbon fibers, on the other hand, have low thermal conductivity and are, therefore, used for thermal insulation purposes. Examples include the nose cone and leading edges of the US Space Shuttle Orbiter. Isotropic pitch-based carbon fibers are also used for insulating purposes because of their low thermal conductivity.

Surface treatments of carbon fibers include oxidative treatments. This results in cleaning of the surface impurities and debris, formation of chemical surface groups (largely acidic) at the carbon fiber surface and a rougher surface morphology, all of which result in enhanced mechanical interlocking with the polymeric matrix.

Unlike most other nonmetallic fibers, carbon fiber can be an electrical and thermal conductor. Medical equipment manufacturers such as 3M, Contour and Conmed use carbon fiber because it is radiotranslucent. This enables its use for lead wires primarily with neonatal electrodes so that hospital personnel do not have to make a decision whether what they see is a copper wire or a vessel, vein or artery. The insulated carbon fiber cannot be seen or mistaken for something else during X-ray and magnetic resonance imaging (MRI) procedures. Carbon fiber conductors have replaced copper wires at an equal cost, but with highly improved performance (Wagner, 1996). Minnesota Wire and Cable Co. produces insulated carbon fibers for both the medical industry and the aerospace industry. Figure 8.15 shows some of these electrode leads made of flexible, radiotranslucent carbon fiber. In the aerospace field, companies such as Ball

Aerospace and Technologies Corporation use carbon fiber to improve the weight and size of other more bulky heat transfer devices, while maintaining suitable thermal conductivity. Insulated carbon fiber can also be shaped and bent to fit the area involved while meeting the heat transfer requirements.

Chapter 9

Experimental determination of fiber properties

Experimental determination of the properties of any material is very important. This is particularly true in the case of fibrous materials because generally insufficient data are available for them. Fibers have one very long dimension and the other two extremely small. This makes determination of their properties, physical and mechanical, far from trivial. In particular, determination of their transverse properties, i.e. in the direction of the fiber diameter, can be difficult. In this chapter we describe experimental techniques to determine some physical and mechanical properties of fibers.

9.1   Physical properties

Many simple physical attributes of fibers can be quite important. We describe some of these attributes and the methods used to measure them.

Weight per unit length. This is also known as the yield and can be measured very simply by weighing a known length of fiber and expressing the result in grams per meter. This parameter is useful in comparing the yield of the final fiber from the precursor fiber, say carbon fiber from the polyacrylonitrile (PAN) precursor fiber.

Fiber diameter. This a very important parameter in characterizing a fiber. One can make a direct measurement of fiber diameter by means of an optical or scanning electron microscope. There is an ASTM standard (D 578) for this purpose.

The main problem with direct measurement is that fiber diameter may not be uniform along the length. An indirect method that gives an average fiber diameter is to weigh a known length of fiber and use the following simple relationship:

$$d = \sqrt{(w / \pi \rho \ell n)}$$

where d is the fiber diameter, w is the mass of the fiber sample, ℓ is the fiber length, ρ is the fiber density, and n is the number of filaments per tow, if a tow of fiber is being used.

Often fibers have an irregular cross-section. The above method will give an equivalent diameter of a fiber having an irregular cross-section. One can also take a photograph of such a fiber and measure the area planimetrically.

Fiber diameter can also be measured by laser beam diffraction (Haege and Bunsell, 1988). The technique can be rapid and systematic. The radius of the fiber is given by

$$r = \lambda s / d$$

where λ is the wavelength of the laser beam, s is the distance between the fiber and the projection screen, and d is the distance between the first intensity minima of the diffraction pattern.

Fiber density. The density of a fiber can be measured by the same techniques as the ones used for conventional, bulk materials. The Archimedes method can be used in a straight forward manner to measure this simple but important property. One weighs the mass of a fiber sample in air and in a liquid of known density, for example, ethanol. Then the following expression gives the density:

$$\rho_f = [m_a / (m_a - m_\ell)] \rho_\ell$$

where ρ_f is the density of the fiber, m_a is the mass of the fiber sample in air, m_ℓ is the mass of the fiber sample in the liquid and ρ_ℓ is the density of the liquid.

Swelling. Absorption of moisture can cause swelling of fibers (see Chapter 2). The change in dimensions caused by swelling can be measured by some of the techniques mentioned above for measuring the diameter of the fiber. Microscopic techniques may be used to measure the diameter of the swollen fiber or the cross-sectional area may be measured by a planimeter. In the case of an irregular cross-sectional fiber, it is better to measure change in area due to swelling than change in diameter. Optical and scanning electron microscopic techniques can be used profitably.

Fabric weave. Fabric weave pattern can influence the properties such as drape-ability, tightness, etc. of a fiber product. The quantities of relevance in this regard are pick count, fabric areal weight, etc.

Thermal analysis. Thermal analysis of fibers is very important. The term thermal analysis represents a group of techniques in which the property of a sample is measured while the sample is subjected to a controlled temperature program. The most common thermal analysis techniques are:

- thermogravimetry (TG) or thermogravimetric analysis (TGA);
- differential scanning calorimetry (DSC);
- thermomechanical analysis (TMA);
- dynamic mechanical analysis (DMA).

Thermogravimetry (TG) or thermogravimetric analysis (TGA). In this technique, the mass of a sample is followed as a function of temperature or time. The amount and rate of mass change with temperature or time in a controlled atmosphere are obtained. Such information can tell us about thermal stability as well as the compositional profile of a variety of elastomers and polymers. It is an excellent quantitative technique but qualitatively there may be some doubt as to what material is lost during heating.

Differential scanning calorimetry (DSC). In this case, the heat inflow and outflow from a sample is measured and information on events that involve a change in the heat content is obtained. For example, melting point, glass transition temperature, heat of fusion, phase transformations, etc.

Thermomechanical analysis (TMA). In this technique, information on changes in the size of a sample is obtained , e.g. thermal expansion and coefficient of thermal expansion, cure shrinkage, glass transition, thermal relaxations, any phase transformation involving volume change in the material. We describe the measurement of the coefficient of thermal expansion in detail later in this section.

Dynamic mechanical analysis (DMA). This technique is mainly used for determining the viscoelastic properties of a sample. The sample is subjected to an oscillating deformation and the amount of energy stored or lost is measured. In a purely elastic material, Hooke's law will be obeyed and the stress and strain will be in-phase. In a viscoelastic material, the ratio of the viscous (or dissipating) energy to elastic (or storage) energy is obtained as tan δ.

Thermal conductivity. The heat flow in a material is proportional to the temperature gradient, and the constant of proportionality is called the thermal conductivity. Thus, in the most general form, using indicial notation, we can write

$$q_i = -k_{ij}\, dT/dx_j$$

where q_i is the heat flux along the x_i axis, dT/dx_j is the temperature gradient across a surface that is perpendicular to the x_j axis, and k_{ij} is the thermal conductivity. As should be evident from the two indexes, thermal conductivity is also a second-rank tensor. Although, k_{ij} is not a symmetric tensor in the most general case, it is a symmetric tensor for most crystal systems. For an isotropic material, k_{ij} reduces to a scalar number, k. For a fiber, we have two constants, longitudinal along the axis and transverse along any direction in the cross-sectional plane, k_1 and k_2, or k_L and k_T, respectively. The same holds true for a transversely isotropic material such as a unidirectionally reinforced fibrous composite, there will two constants: thermal conductivity in the axial direction, k_{cl}, and that in the transverse direction, k_{ct}. Most thermal conductivity data on fibers are obtained indirectly from measurements made on unidirectionally reinforced fibrous composites. The thermal conductivity in the axial direction, k_{cl} can be predicted by a rule-of-mixtures or action-in-parallel-type expression

$$k_1 = k_{cl} = k_{fl}\, V_f + k_m\, V_m$$

where k_{fl} is the thermal conductivity of the fiber in the axial direction and k_m is that of the isotropic matrix, and V_f and V_m are the volume fractions of the fiber and the matrix, respectively.

In a transverse direction, the thermal conductivity of a unidirectionally aligned fiber composite (i.e. transversely isotropic) can be approximated by the action-in-series model. This would give

$$k_{ct} = k_2 = k_{f2}\, k_m\, /(k_{f2}\, V_f + k_m\, V_m)$$

Knowing all the parameters except the thermal conductivity of the fiber, we can use these expressions to obtain thermal conductivity. It should be noted that in real composites, it is likely that the thermal contact at the fiber/matrix interface will be less than perfect because of thermal mismatch between the fiber and the matrix.

A laser flash technique can be used for determining the thermal conductivity of fibers (Whittaker et al., 1990). The laser flash technique provides us the diffusivity, a, which, in turn, is a function of the thermal conductivity, specific heat, and density of the material. Thus,

$$\Delta T = f\left(\frac{1}{a}\right) = f\left(\frac{C_p \rho}{k}\right)$$

where C_p is the specific heat, ρ is the density, and k is the thermal conductivity. The laser flash technique measures the temperature change with time that gives the value of diffusivity, a. A small disk-shaped sample is subjected to a burst of radiant energy. Under the assumption that this energy pulse is absorbed instantaneously and uniformly, the thermal diffusivity can be calculated from the following relationship (Whittaker *et al.*, 1990).:

$$a = 0.1388 \, d^2/t_{0.5}$$

where d is the sample thickness and $t_{0.5}$ is the half-risetime of the energy pulse.

If the specific heat of a single fiber is very small (e.g. carbon fiber), then, the thermal conductivity measurements should be made on a sample containing many fibers.

Coefficient of thermal expansion. Much of the data on thermal expansion behavior of fibers has been obtained indirectly from oriented fiber reinforced composites. In the case of composites such as carbon fibers in a resin matrix, the longitudinal coefficient of thermal expansion will be dominated by the carbon fibers because the axial coefficient of thermal expansion of carbon fibers is almost zero. In the transverse direction, however, the matrix contribution to coefficient of thermal expansion can be quite significant. Sheaffer (1987) used a laser diffraction technique to determine the coefficient of thermal expansion of carbon fibers in the transverse direction. The fibers were held under a tension of $7 \, \text{kPa}$. Not unexpectedly, the transverse expansion coefficient of carbon fibers ranged between $6.7 \times 10^{-6} \text{K}^{-1}$ and $13.1 \times 10^{-6} \text{K}^{-1}$. Tsai and Daniels (1994) used a length of fiber fixed at the ends on a plate of titanium silicate (which has a low coefficient of thermal expansion of $3 \times 10^{-6} \text{K}^{-1}$) inside an oven. The fiber was loaded at its center by small incremental weights and the deflections were obtained photographically. From the applied loads and the corresponding deflections, the authors obtained the elastic stiffness of the fiber. For determining the coefficient of thermal expansion, the authors obtained thermal strains from the deflections at a constant load, but varying temperatures.

Electrical resistivity. Electrical resistivity (or conductivity) of a fiber can be measured by means of simple resistance probe. The fibers are hung by a weight to keep them taut. Electrical resistance is measured at specific intervals of length. The slope of a plot of electrical resistance against length gives the resistivity. The following expression gives the electrical resistivity, ρ:

$$\rho = \pi d^2 R / 4\ell$$

where d is the fiber diameter, R is the electrical resistance, and ℓ is the fiber length.

Wettability or contact angle. Wettability is a very general term and can be defined as the ability of a liquid to spread on a solid surface. Consider a liquid drop

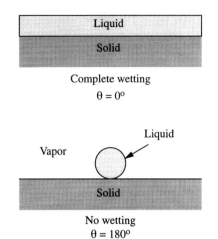

Figure 9.1 Contact angle as a measure of wettability. For $\theta=0°$, we have perfect wetting and the liquid will spread spontaneously on the surface in the form a thin film. For $\theta=180°$, the liquid will ball up, and there will be no wetting. In between these two extremes, for $0°<\theta<180°$, we have partial wetting.

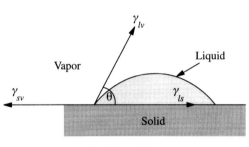

resting on a solid substrate and the equilibrium of forces in this system. The three forces are the three specific energies or surface tensions (i.e. energy per unit area): γ_{sv}, the energy of the solid/vapor interface; $\gamma_{s\ell}$, the energy of the liquid/solid interface; and $\gamma_{\ell v}$, the energy of the liquid/vapor interface, see Fig. 9.1. When we put a liquid drop on a solid substrate, we replace a portion of the solid/vapor interface by a liquid/solid and a liquid/vapor interface. Thermodynamically, spreading of the liquid will occur only if this results in a decrease in the free energy of the system, i.e.

$$\gamma_{s\ell}+\gamma_{\ell v}<\gamma_{sv}$$

The most general situation of partial wetting is shown in Fig. 9.1. From the equilibrium of forces in this figure, we have;

$$\gamma_{s\ell}+\gamma_{\ell v}\cos\theta<\gamma_{sv}$$

where θ is the contact angle, so that

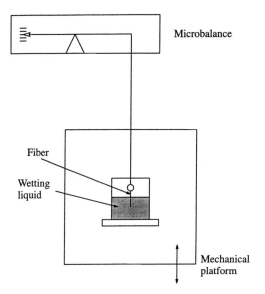

Microbalance

Fiber

Wetting
liquid

Mechanical
platform

Figure 9.2 Measurement of contact angle by the Wilhelmy technique. The wetting force is measured by a sensitive electrobalance and the contact angle is calculated from changes in the wetting force.

$$\theta = \cos^{-1}\left[\frac{\gamma_{sv} - \gamma_{s\ell}}{\gamma_{\ell v}}\right]$$

This contact angle is taken as a measure of wettability of a solid, in the present case, a fiber. For $\theta = 0°$, we have perfect wetting and the liquid will spread spontaneously on the surface in the form a thin film. For $\theta = 180°$, the liquid will ball up, and there will be no wetting. In between these two extremes, for $0° < q < 180°$, we have partial wetting. The contact angle for a given system can vary with temperature, hold time, presence of any adsorbed gases, etc. The contact angle of a fiber in a given liquid is an important parameter for a variety of applications. For example, such information is very useful for dyeing a fiber or, in the case of a composite, it is useful for finding out if a particular matrix material in the liquid form will wet the fiber. Many techniques are available for measuring the contact angle between a liquid and a fiber (see, for example, Schultz and Nardin, 1992; Bascom, 1992). An important technique for obtaining the contact angle, called the Wilhelmy or tensiometric technique, involves a measurement of the wetting force by means of a sensitive electrobalance and a mechanical elevator, as shown in Fig. 9.2. The contact angle is calculated from changes in the wetting force. The reversible elevator or the mechanical platform, shown in Fig. 9.2, is driven by two hydraulic cylinders in the vertical direction. This allows the measurement of advancing and receding angles of contact. The apparatus is mounted on a vibration damping table. A single fiber, suspended from one arm of an electromechanical balance, is immersed in a beaker containing the liquid of interest. The force acting on the fiber when it touches the surface of the liquid and when it penetrates the liquid is a function of the following variables:

- fiber surface morphology;
- density of the liquid;
- depth of immersion;
- contact angle of the liquid on the fiber surface.

The following expression relates these parameters (Bascom, 1992)

$$F = 2\pi r \gamma_{lv} \cos\theta - \rho\pi r^2 gh \qquad (9.2)$$

where r is the fiber radius, θ is the contact angle, γ_{lv} is the surface energy of the liquid/vapor interface, ρ, is the density of the liquid, h is the depth of immersion, and g is the acceleration due to gravity. The first term on the right-hand side of Eq. (9.2) is the capillary force and the second term is the buoyant weight of the submerged fiber. Interestingly, for very fine diameter fibers ($<50\,\mu m$), the buoyancy term becomes negligible because the radius term is squared. Consequently, Eq. (9.2) reduces to

$$F = 2\pi r \gamma_{lv} \cos\theta \qquad (9.3)$$

If the force is mg, m being the mass of the fiber, then using $2r = d$, the fiber diameter and rearranging Eq. (9.2), we get the following expression for the contact angle:

$$\cos\theta = mg/\pi d\,\gamma_{lv} \qquad (9.4)$$

Note that we need the fiber diameter as input in Eq. (9.4). A quick and convenient method to do this is to use a liquid that has a $\theta = 0°$, then use the following relationship to determine the diameter, d:

$$d = mg/\pi\gamma_{lv}$$

Dynamic contact angle. This involves the measurement of contact angle while immersing (advancing contact angle) a fiber into a liquid and pulling it out (receding contact angle) from a liquid. The dynamic contact angle is measured continuously while the fiber is immersed or pulled out. The continuous traces recorded are the tensiometer weight versus the distance along the fiber; the weight can be converted to contact angle via Eq. (9.4) to obtain a plot of contact angle versus the distance along the fiber. An important phenomenon that is frequently observed in such experiments is called *contact angle hysteresis*, which is a lag in the contact angle of the liquid drop on the solid surface when the fiber is immersed and when the fiber is pulled out of the liquid (Neumann and Good, 1979)

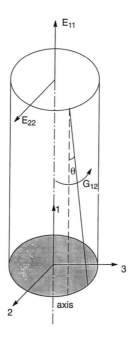

Figure 9.3 Elastic constants of an anisotropic fiber: the longitudinal Young's modulus of fiber, E_{11} or E_L, the transverse Young's modulus E_{22} or E_T, and the principal shear modulus, G_{12} or G_{LT}. Not shown are the two Poisson's ratios: ν_{12} or ν_{LT}, the longitudinal Poisson's ratio of the fiber and ν_{21} or ν_{TL}, the transverse or in-plane Poisson's ratio of the fiber cross-section.

9.2 Mechanical properties

Determining mechanical characteristics of fibrous materials is far from simple, mainly because of their small diameter. In particular, in the case of anisotropic fibers such as carbon or aramid, we need to determine five elastic constants, assuming isotropy in the cross-sectional plane. Figure 9.3 shows three of the five elastic constants: the longitudinal Young's modulus of fiber, E_{11} or E_L, the transverse Young's modulus E_{22} or E_T, and the principal shear modulus, G_{12} or G_{LT}. Not shown are the two Poisson ratios: ν_{LT}, the longitudinal Poisson's ratio of fiber and ν_{TL}, the transverse or in-plane Poisson's ratio of fiber cross-section. We provide below some salient features of various techniques to determine such characteristics.

9.2.1 Longitudinal tensile strength

Tensile testing of individual filaments is frequently done. It is a tedious process because of the fineness of the fibers and, frequently, because of their brittleness. There is an ASTM standard (ASTM-3379-75) for specimen preparation and testing high modulus (>20 GPa) single filaments. Other ASTM standards available for measuring the strength of fibers include: D3479-76, D3800-79, D3878-87, D4018-81. There are certain critical requirements for testing fibers.

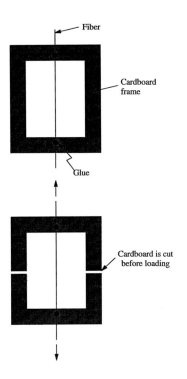

Fiber

Cardboard frame

Glue

Cardboard is cut before loading

Figure 9.4 The cardboard frame technique to measure the tensile strength of a single filament. The single filament is adhesively bonded to a cardboard frame with a longitudinal slot. The cardboard, with the bonded fiber, is clamped in the grips of a tensile testing machine. The cardboard frame is cut off on the two sides, leaving the single fiber ready to be tested.

Alignment of fiber is an obvious one. A serious problem with some of the high strength and stiffness fibers is the likelihood of breaking in the grips. Clearly, the fiber under test should not be damaged by the gripping mechanism. In order to avoid such premature failure, some kind of facing material, polyethylene or neoprene, is generally placed between the filament and the metallic grips. This technique is fairly commonly used. As examples, we may cite the work of McMahon (1973); Jain and Abhiraman (1987); and Jain *et al.* (1987) who used this technique to evaluate the properties of carbon fibers while Xu *et al.* (1993) used this technique to study aluminosilicate fibers. The general setup is shown in Fig. 9.4. The single filament is adhesively bonded to a cardboard frame with a longitudinal slot. The cardboard, with the bonded fiber, is clamped in the grips of a tensile testing machine. After all the adjustments and alignment procedures are done, the cardboard frame is cut off on the two sides, leaving the single fiber ready to be tested. Now when the load is applied, only the fiber experiences the load. The specimen is pulled to failure, and a record of load and *apparent* deformation is obtained. The tensile strength is the maximum load divided by the cross-sectional area of the fiber. The precise deformation in the fiber length, however, is not so easy to obtain. One needs to subtract the machine compliance from the total compliance to get the true fiber deformation. But, before we describe this indirect method of obtaining the strain in the fiber, we should point out that although direct measurement of strain in the gage length of a fiber is not easy, it can be done. We first describe how direct measurement of strain in the fiber gage length can be done by laser extensometry and then describe the indirect procedure.

Extensometer is a general term for a device that measures change in length (extension or contraction) of a sample. Commonly, extensometers require mechanical attachment to the sample being measured. Such a procedure will not work with very fine fibers. Laser extensometry provides a noncontact measurement of elongation. The technique measures axial strain by monitoring surface displacement of two points on the specimen. Two laser beams that differ in frequency by a precise amount, say 250 kHz, are brought together at the sample surface at a fixed angle. The location of any two points on the specimen is defined by the point where recombination of the laser beams occurs. Now, when a specimen is strained, relative movement of two points on the specimen surface occurs. This results in a phase change in the recombined beam which can be used as a measure of strain in the specimen. Motion is simultaneously measured at two locations. Interference fringes form where the two beams overlap. The fringe spacing d is given by

$$d=\frac{\lambda}{2\sin\theta}$$

where λ is the wavelength of the laser beam and θ is the half angle between the two incident laser beams. Because of the different frequencies of the two beams, the fringes move relative to the stationary sample at a rate equal to the difference frequency, $\Delta\nu=(\nu_1-\nu_2)$. Laser extensometry has been used to measure strain in a tungsten filament at 2700°C and in 8 μm carbon fiber.

An indirect of way of avoiding the measurement of gage length deformation is as follows (Nunes and Klein, 1967). We can write for the total cross-head deflection, δ_t ,as follows:

$$\delta_t=\delta_n+\delta_c+\delta_m \tag{9.5}$$

where δ_n is the deflection due to the nominal gage length of the fiber ℓ_n, δ_c is the deflection due to the length of the fiber in the jaws, and δ_m is the deflection due to the load cell and the testing machine. We can reasonably assume that the terms, δ_c and δ_m are constants at a given load and independent of the nominal gage length. Also, one can use the relationship

$$\delta_t=\epsilon\ell_n \tag{9.6}$$

where ϵ represents strain. Equation (9.6) allows one to rewrite Eq. (9.5) as

$$\delta_t=\epsilon\ell_n+c \tag{9.7}$$

where c is a constant. Thus, at a constant load, if we plot δ_t against ℓ_n, we should get a straight line whose slope is the strain, ϵ. If we measure these deflections of different fiber lengths at a constant load in the elastic regime, we get the true elastic strain in the fiber at that load, from which we can obtain the elastic modulus.

We should mention that for textile fibers, there is an ASTM D2101 standard which allows fibers to be gripped directly because such fibers are generally not damaged by the clamp pressure. In general, the clamps are lined with rubber or elastomeric pads to reduce the friction.

We make a brief mention of some other problems in testing single filaments. The problem is even more complex for whiskers because of their extremely small dimensions (Kelsey, 1970). Marsh (1961) used an adhesive cement to mount the whiskers on quartz rods. Bending moments will be introduced if a fiber of a whisker is not properly aligned. Petrovic *et al.* (1985) describe the complicated and very careful procedure they developed to test silicon carbide whiskers. A whisker was glued to the beveled edge of a glass microscope slide with a fast curing epoxy. Silicon carbide powder was added to the glue to increase its modulus. After performing very careful alignment of the whisker, the half glued whisker was examined in SEM for any damage, diameter changes, etc. After the SEM examination, the glass slide with the whisker was clamped on to the linear displacement device of the tensile tester. A second glass slide was mounted on an X-Y-Z stage and positioned under the free end of the whisker. The whisker was then aligned using an optical microscope. The second glue joint was made after the alignment. The whisker was then loaded to failure and the load–displacement curve recorded.

So far we have discussed the measurement of mechanical properties at room temperature. Obtaining mechanical property data at high temperature is even more difficult. One needs such data, especially the strength of ceramic fibers at the high temperatures at which they are likely to be used. A major problem comes from the thermal gradients involved between the gage length and the portion in the grips. Unal and Lagerlof (1993) tested alumina fibers to 1500°C by pulling the entire fiber and grips in the hot zone of the furnaces, thus avoiding thermal gradients associated with cold grips, i.e. outside the hot zone. Beads of an alumina adhesive were deposited on the alumina fiber extremities with much skill and patience.

9.2.2   Recoil compression test

Obtaining the compressive strength of a slender fiber is even more difficult than measuring its tensile strength. An indirect technique has been devised to obtain compressive strength of fibers. It is well known that the recoil of a tensile stress wave generates a compressive wave. This principle is used in the recoil compression test, which allows one to estimate the compressive strength of fibers.

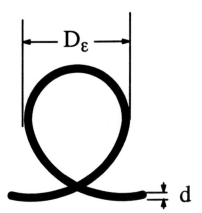

Figure 9.5 A knot test. D_ϵ, the loop diameter, is approximately equal to d/ϵ where d is the fiber diameter and ϵ is the failure strain.

A rigidly held fiber is stretched in tension to a certain stress and cut at the mid-point by a pair of scissors or an electric spark. The tensile wave is reflected at the rigid clamps into a compressive wave.

9.2.3   Loop test

Sinclair (1950) devised this loop test to estimate the strength and elastic modulus of fibers. There are many variants of this test, which is basically a kind of bend test. Greenwood and Rose (1974), among others, used such a test to evaluate Kevlar aramid fiber. Hüttinger (1990) used such a loop or knot test to determine the deformation characteristics of carbon fibers. Figure 9.5 shows the knot test schematically. If d is the fiber diameter and D_ϵ is the loop diameter corresponding to a tensile strain of ϵ in fiber, then we can write

$$D_\epsilon \approx d/\epsilon = Ed/\sigma$$

where σ is the stress and E is Young's modulus. As in any bending beam test, the stresses will be higher at the fiber surface and decrease toward the axis of the fiber. Clearly, in a fiber having a skin–core structure, such a test will be biased toward the characteristics of the skin region. Hillig (1981) used a similar setup to test fibers, see Fig. 9.6. A fiber, lying on a rubber mat base, is bent by a downward force applied to a cylindrical tool. Siemers *et al.* (1988) tested SCS-6 silicon carbide fiber by looping around a series of successively smaller drill bits until the fiber failed. The failure strain e was computed from the relationship,

$$\epsilon = d/D$$

where d is diameter of the monofilament and D is the diameter of the drill bit. Since the fiber is elastic to fracture, one can compute the failure stress from

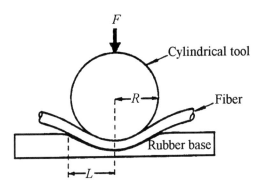

Figure 9.6 A fiber, lying on a rubber mat base, is bent by a downward force applied to a cylindrical tool (after Hillig, 1981).

Hooke's law, $\sigma = E\epsilon$, where E is the Young's modulus of the fiber. The failure strain was taken as the average of the strain imposed by the drill bit that caused fiber failure and the one immediately before that.

9.2.4 Dynamic elastic modulus

Speed of sound in a material is related to its Young's modulus and density by the following relationship

$$c^2 = E/\rho$$

A transducer is used to send a pulse down the fiber while another transducer set at a fixed distance from the first one detects the arriving pulse. Smith (1972) used such an ultrasonic technique to measure the Young's modulus of carbon fibers.

The modulus measured by making an acoustic pulse travel down a fiber axis is called the dynamic modulus to distinguish it from the static modulus measured in a mechanical testing machine.

9.2.5 Shear modulus

Fibers are frequently twisted in a variety of operations. Thus, it is important to know the torsional or shear modulus of a fiber. In Fig. 9.7 a cylindrical fiber being twisted under the action of a torque shows the shear modulus corresponding to a torsion angle of θ. The shear strain, γ, is given by

$$\gamma = \theta r/\ell \tag{9.8}$$

while the shear stress, τ, is related to shear strain via the shear modulus, G:

$$\tau = G\,\theta r/\ell \tag{9.9}$$

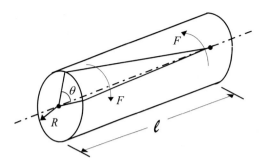

Figure 9.7   A cylindrical fiber of length ℓ subjected to torque.

where r is the fiber radius, ℓ is the fiber length, and θ is the angle of twist. Let us consider the torsion of the cylindrical fiber, Fig. 9.7. We apply a torque to the fiber by two opposing but equal forces, F at the two ends. Let da be the area over which a small force dF is acting, then

$$\tau = dF/da$$

This elemental area is given by $2\pi r dr$, where dr is the thickness of the annular area on which the force dF is acting. Thus,

$$\tau = dF/2\pi r\ dr \tag{9.10}$$

From Eqs. (9.9) and (9.10), we can write

$$G\theta r/\ell = dF/2\pi r\ dr$$

$$dF = 2\pi r^2\ G\theta dr/\ell \tag{9.11}$$

The torque at a radius r from the central axis is rdF. The total torque is obtained by integrating from 0 to R, the fiber radius. Thus

$$\text{total torque} = \int r\ dF = \delta \int_0^R 2\pi r^3 G\ (\theta/\ell)\ dr = (\pi R^4 G\theta)/2\ell \tag{9.12}$$

The quantity $\pi R^4 G/2$ is called the torsional rigidity which is a combined function of the shear modulus, G and a geometric term containing the fiber radius to the power four! Thus halving the fiber diameter will result in reducing the torsional rigidity by a factor of 16.

One can determine the shear modulus of a fiber from a torque per unit area versus twist curve. In practice, a simple apparatus called a torsion pendulum is used more commonly. An experimental setup to measure the shear modulus of small fibers is shown in Fig. 9.8 (Mehta, 1996). The torsional pendulum, placed in a vacuum oven, allows the measurement of shear modulus as a func-

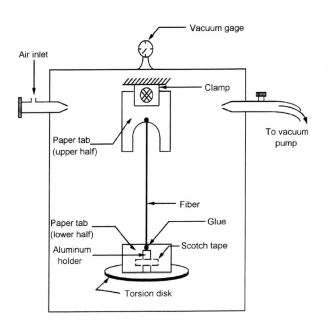

Figure 9.8 An experimental setup to measure shear modulus of small fibers (after Mehta, 1996).

tion of temperature under vacuum. In this particular example, the fiber sample was mounted on a cardboard disk. An aluminum fastener was fixed at the center of the cardboard to hold the paper tab on which the fiber sample was mounted. After mounting the paper tab and sample, the top end was fixed on a clamp stand inside the oven. The two sides of mounted paper tab were then cut and only the fiber was freely attached to the torsional pendulum. A small initial twist is given to the fiber and the amplitude of the successive oscillations is determined. The shear modulus of a fiber of a circular cross-section is given by

$$G = 8\pi I \ell / T^2 R^4$$

where I is the moment of inertia of the torsion pendulum, ℓ is the fiber length, R is the fiber radius and T is the period of oscillations. The period of oscillations is given by

$$T = T_0 / \sqrt{(1 + (\Delta/2\pi)^2}$$

where T_0 is the experimentally measured period of oscillation, Δ is the logarithmic decrement $= \ln(A_2/A_1)$, where A_1 and A_2 are the amplitudes of two successive oscillations. Note that the radius term has an exponent of 4, which means that even a very small error in the measurement of fiber radius will introduce a significant error in the shear modulus value.

9.2.6   Transverse properties

Many high performance fibers such as aramid, polyethylene, carbon, etc. can have highly anisotropic properties. For example, the elastic modulus, the strength, the coefficient of thermal expansion, thermal conductivity, etc. may be quite different along the fiber axis and transverse to the fiber axis. This anisotropy stems from the highly oriented chain structure in the case of organic fibers, from the layer structure in the case of carbon and boron carbide, or from the oriented structure obtained by CVD such as is the case for silicon carbide or boron fibers. In the case of pure single crystal fibers such as alumina, the anisotropy will have origin in the crystal structure. Hadley *et al.* (1969) measured the transverse modulus of polyethylene and some other monofilaments in the rod form. They compressed the thick rodlike specimen of fibers in the transverse direction between two parallel glass plates and measured, with a microscope, the flattening of the contact area of the fiber. The transverse modulus was obtained from the force applied and the size of the contact area. This method is suitable for large diameter fiber. Phoenix and Skelton (1974) used an Instron tester fitted with two parallel steel platens to study the transverse compressive behavior of aramid, nylon, polyester and carbon fibers. Jawad and Ward (1978) performed transverse compression tests on nylon 6.6 and linear polyethylene samples machined from extruded rods (2.5 mm diameter). Thus, most attempts at measuring the transverse modulus of polyethylene and other polymeric fibers have involved compressing the single filament between two parallel plates and measuring the amount of flattening of the contact area of the fiber under compression. Eldridge *et al.* (1993) measured transverse strength of SCS-6 silicon carbide fiber by a similar technique. Others used the indirect route, namely testing a fibrous composite and obtaining an estimate of the transverse modulus of the fiber from the transverse modulus of a unidirectionally reinforced fiber/matrix composite. The assumption here is that the in situ behavior of the fiber is the same as that outside the composite. Kawabata (1990) devised a technique to measure the transverse modulus of fibers around $10 \, \mu m$ in diameter by recording compressional force and diametral change at a constant loading rate. The bottom plane was a mirror-finished, flat steel bed while the top plane was a square plane, mirror-finished. A single fiber was placed horizontally on the bottom plane and compressed by the top plane, which was driven by a rod connected to an electromagnetic driver (load capacity=50 N). A transducer was used to measure the force and a linear differential variable transformer (LVDT) was used to measure the deformation of the fiber in the diametral direction (resolution $\sim 0.05 \, \mu m$). An experimental load-deflection curve was obtained and compared with the load-deflection curve obtained from a theoretical model, assuming isotropy in the cross-sectional plane. The change in diameter, u, as a function of compressive force, F, can be obtained from the following relationships:

$$u=(4F/\pi) \{(1/E_T)-(v^2_{LT}/E_L)\} \{0.19+\sinh^{-1}(R/b)\}$$

$$b^2=(4FR/\pi)\{(1/E_T)-(v^2_{LT}/E_L)\}$$

where u is the change in fiber diameter, b is half the contact width between the fiber and the punch face, F is the compressive force per unit length of fiber, E_L is the longitudinal modulus of the fiber, E_T is the transverse modulus of the fiber, R is the radius of the fiber, and v_{LT} is the longitudinal Poisson's ratio of the fiber (defined by the longitudinal strain and the transverse contraction strain when the fiber is uniaxially stretched along its axial direction). As can be seen, the transverse modulus, E_T, is an input parameter here. This E_T is used as a curve fitting parameter to match the experimental curve. Note also that the Poisson's ratio term, v, enters as a squared term. Poisson's ratio being very small, the square of Poisson's ratio will be an insignificant quantity and can be neglected in most cases. Kawabata's (1990) measurements on carbon and aramid fibers showed them to be highly anisotropic. He observed a correlation between transverse and longitudinal moduli. Carbon and other ceramic fibers failed in a brittle manner and E_T decreased sharply with increasing E_l. Aramid and polyethylene UHMW PE failed in a ductile manner. Organic fibers showed a small increase in E_T with increasing E_L. Jones et al. (1997) studied the transverse compression deformation of a number of commercially available organic and inorganic fibers as well as that of thermally cross-linkable poly(p-1,2-dihydrocyclobutaphenylene terephthalamide) (PPXTA) fiber. The apparatus used by these authors had a stationary flat-ended punch and a movable table supporting the fiber. A load–displacement curve was obtained was obtained in each case. Carbon and alumina fibers showed a brittle behavior, while organic fibers showed a nonlinear deformation.

9.2.7 Micro-Raman strain measurement

A technique called micro-Raman strain measurement, has become quite important in measuring strain in nonmetallic fibers. Raman spectroscopy deals with the phenomenon of a change in frequency when light is scattered by molecules. An exchange of energy between the scattering molecule and the photon of the incident radiation gives rise to the Raman effect, named after C.V. Raman who was awarded the Nobel prize for his discovery of this phenomenon. Raman scattering spectroscopy uses a monochromatic source of radiation, generally obtained from a laser. The Raman frequency (or the wavenumber) depends on the spring constant. The interatomic or intermolecular force being nonlinear, the spring constant will change if we constrain the molecule with a bond length different from that it would have in the unconstrained state. This implies that a shift will occur in the Raman frequency if we stress the solid. This phenomenon has been

(a)

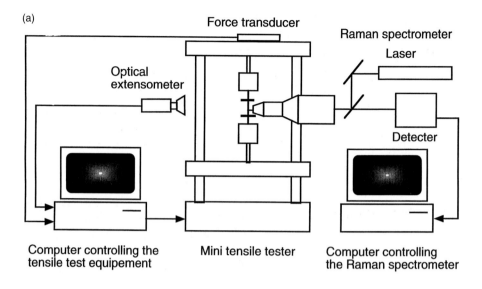

Force transducer

Raman spectrometer

Laser

Optical
extensometer

Detecter

Computer controlling the   Mini tensile tester   Computer controlling
tensile test equipement   the Raman spectrometer

(b)

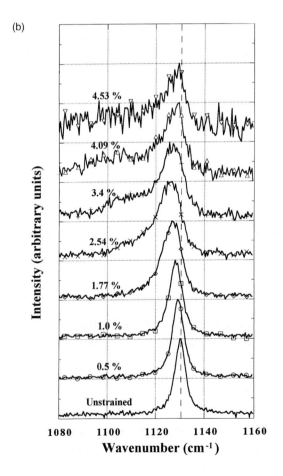

Figure 9.9 (a) An experimental setup to obtain the characteristic Raman spectra from a fiber. (b) The frequency shift for a polyethylene fiber under tension at room temperature. The strain values are indicated. (Courtesy of Berger and Kausch, 1996.)

successfully employed to study the deformation behavior of organic as well as inorganic fibers (Cannon and Mendoza, 1991; Kip *et al.*, 1991; Boogh *et al.*, 1992; Day and Young, 1993; Galiotis, 1991; Wong and Young, 1994; Young *et al.*, 1990; Young, 1996). Raman spectroscopy is a very powerful technique that can provide a significant structural information. Equipment involving the use of a charge coupled device (CCD) detector and more computer control has extended the Raman microprobe into a Raman microscope in which one can obtain images of the Raman scattered light and record the spectra under confocal conditions (Batchelder *et al.*, 1991; Sudiwala *et al.*, 1992; Williams *et al.*, 1994) .

An experimental setup to obtain the characteristic Raman spectra from these fibers is shown in Fig. 9.9a (Berger *et al.*, 1996). Under tension, the peaks of the Raman bands shift to lower frequencies. Figure 9.9b shows such a frequency shift for a polyethylene fiber. It turns out that the $1127 \, cm^{-1}$ Raman band in polyethylene fiber corresponds to symmetric stretching of the C–C bond. When a polyethylene fiber is strained, this peak would shift to lower frequencies. Similar results have been obtained with aramid fibers. There is an approximately linear shift in the peak position, Δv with strain, e. The slope of the line, $d\Delta v/de$, is proportional to the fiber modulus E_f (Young, 1996). This means that one can use the Raman spectroscopy to investigate molecular stretching, since elastic modulus is a measure of molecular bonding. In reality, the magnitude of frequency shift is a function of the material, Raman band under consideration, and the Young's modulus of the material. The shift in Raman bands results from changes in force constants due to changes in molecular or atomic bond lengths, and bond angles. We can write

$$hv_i + E_1 = hv_s + E_2$$

where h is Planck's constant, v_i is the frequency of incident radiation, v_s is the frequency of scattered radiation, and E_1 and E_2 are the initial and final energies of the molecule, respectively. The frequency shift (+ or −) is given by

$$\Delta v = v_s - v_i$$

A set of Raman frequencies of the scattering species forms the Raman spectrum. Raman frequency shift is commonly expressed in term of a wavenumber (cm^{-1}). Thus

$$\text{Raman shift} = (v_i - v_s)/c = (E_2 - E_1)/h$$

where c is the velocity of light. The necessary condition for Raman scattering is that the incident photon energy hv_i must be greater than the energy difference between the final and initial states of actual transition.

Chapter 10

Statistical treatment of fiber strength

Fracture of brittle materials, in general, involves statistical considerations. Materials have randomly distributed defects on their surfaces or in their interior. Fibrous materials, as we saw in Chapter 2, have a large surface area per unit volume. This makes it more likely for them to have surface defects than bulk materials. The presence of defects at random locations can lead to scatter in the experimentally determined strength values of fibers, which calls for a statistical treatment of fiber strength. Clearly, such scatter will be much more pronounced in brittle fibers than in ductile fibers such as metallic filaments. This is because ductile metals will yield plastically rather than fracture at a flaw of a critical size. Thus, most high performance fibers, with the exception of ductile metallic filaments, shows a rather broad distribution of strength because they are highly flaw sensitive. Since the distribution of flaws is of statistical nature, the strength of a fiber must be treated as a statistical variable. To bring home this important point of variation in strength of a fiber as a function of fiber length, we show, in Figs. 10.1–10.3, the variation of tensile strength of some fibers as a function of gage length: high modulus carbon fiber (Fig. 10.1), boron fiber (Fig. 10.2), and Kevlar 49 aramid fiber (Fig. 10.3). Intuitively, one can see that the probability of finding a critical flaw (which corresponds to the failure strength) increases as the volume of brittle material increases. In the case of a fiber, this translates into an increase in the probability of finding a critical flaw as the fiber length increases. Put more succinctly, the average fiber strength decreases with increasing fiber gage length. An increase in fiber diameter can have similar effect. In this chapter, we provide a brief review of the statistical variation of fiber strength.

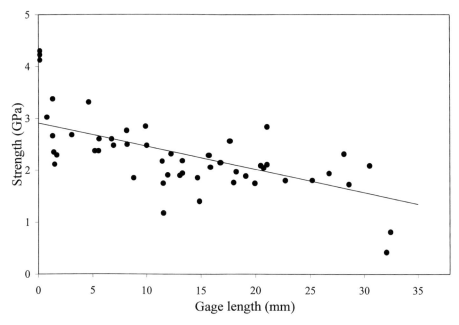

Figure 10.1   Tensile strength of high modulus carbon fiber as a function of gage length (after Diefendorf and Tokarsky, 1975).

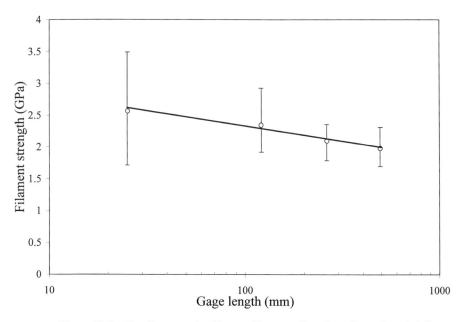

Figure 10.2   Tensile strength of boron fiber as a function of gage length (after Herring, 1965).

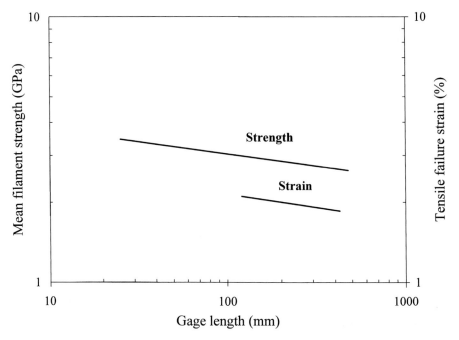

Figure 10.3 Tensile and failure strain of Kevlar 49 aramid fiber as a function of gage length (after Zweben, 1977).

10.1   Variability of fiber strength

We can regard a fiber as consisting of a chain of links. We assume that fiber failure occurs when the weakest link fails. This is called the weakest-link assumption. It turns out that such a weak-link material is well described by the statistical distribution known as the Weibull distribution (Weibull, 1939, 1951). We first describe the general Weibull treatment for brittle materials and then describe its application for fibers.

10.2   Weibull statistics

Let us say that a series of identical samples are tested to failure. From such tests we can obtain the fraction of identical samples, each of volume V_0, that survives when loaded to a given stress, σ. Let us call this $S(V_0)$. According to the Weibull distribution, this survival probability is given by

$$S(V_0)=\exp\left[-\left(\frac{\sigma}{\sigma_0}\right)^{\beta}\right] \tag{10.1}$$

where σ_0 and β are constants. Figure 10.4a shows a general plot of survival probability $S(V_0)$ against strength σ. At $\sigma=0$, i.e. no load on the sample, all samples

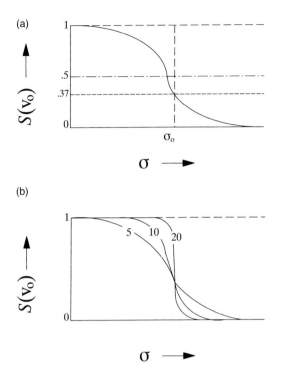

(a)

$S(V_0)$

σ

(b)

$S(V_0)$

σ

Figure 10.4 A plot of survival probability, $S(V_0)$ against strength σ. At $\sigma=0$, i.e. no load on the sample, all samples survive and $S(V_0)=1$. As σ increases, more samples fail and $S(V_0)$ decreases. Eventually, $S(V_0) \rightarrow 0$, as $\sigma \rightarrow \infty$, i.e. all samples fail at very high loads. (b) Survival probability versus strength for different values of Weibull modulus, β. As β increases, the distribution becomes less broad.

survive and $S(V_0)=1$. As σ increases, more samples fail and $S(V_0)$ decreases. Eventually, $S(V_0) \rightarrow 0$, as $\sigma \rightarrow \infty$, i.e. all samples fail at very high loads. It is easy to see from Eq. (10.1) that for $\sigma = \sigma_0$,

$$S(V_0) = \frac{1}{e} = 0.37$$

This expression tells us that the constant, σ_0 is the stress corresponding to a survival probability of 37%. β is called the Weibull modulus and is a measure of the flaw distribution in the sample. The smaller the value of β, the greater the *variability* in strength. In other words, the value of Weibull modulus β tells us how rapidly the strength falls as we approach σ_0. Figure 10.4b shows plots of survival probability versus strength for different values of Weibull modulus β. As β increases, the distribution becomes less broad. For a given material, by using an improved processing technique, one can make a flaw size distribution less broad and thus increase the value of β. Table 10.1 gives some typical β values for bulk materials. In general, brittle materials have a lower Weibull modulus than ductile materials.

If we take logs on both sides of Eq. (10.1), we obtain

$$\ln \left[\frac{1}{S(V_0)} \right] = \left(\frac{\sigma}{\sigma_0} \right)^{\beta}$$

Taking logs again, we have

$$\ln \ln \left[\frac{1}{S(V_0)} \right] = \beta \ln \left(\frac{\sigma}{\sigma_0} \right)$$

We show in Section 10.5 how the Weibull modulus can be obtained graphically by using a double log plot of such an equation that gives a straight line of slope β.

10.3   Weibull distribution of fiber strength

The Weibull distribution is called a parametric distribution, i.e. it is an empirical distribution and does not concern itself with the origin of the defects. The Weibull distribution for the strength (σ) of a brittle material takes the following form

$$f(\sigma) = L\alpha\beta\sigma^{\beta-1} \exp(-L\alpha\sigma^{\beta}) \tag{10.2}$$

where L is the fiber length, σ is the fiber strength, and α and β are statistical parameters. $f(\sigma)$ is a probability density function.

We define the kth moment, M_k, of a statistical distribution, as

$$M_k = \int_0^\infty \sigma^k f(\sigma) \, d\sigma \tag{10.3}$$

Consider a fiber that follows the strength distribution given by Eq. (10.2). We can obtain an expression for the mean strength as follows:

$$\bar{\sigma} = M_1 = \int_0^\infty \sigma f(\sigma) \, d\sigma \tag{10.4}$$

$$= \int_0^\infty L\alpha\beta\sigma^{\beta-1}\sigma \exp(-L\alpha\sigma^{\beta}) \, d\sigma$$

Knowing $d(L\alpha\sigma^{\beta}) = L\alpha \beta\sigma^{\beta-1} \, d\sigma$, the above expression becomes

$$= \int_0^\infty \exp(-L\alpha\sigma^{\beta}) \frac{(L\alpha\sigma^{\beta})^{1/\beta}}{(L\alpha)^{1/\beta}} d(L\alpha\sigma^{\beta}) \tag{10.5}$$

Let x denote $L\alpha\sigma^{\beta}$. Then Eq. (10.5) becomes

Table 10.1 *Typical Weibull modulus (β) values for bulk materials.*

Fiber material	Weibull modulus, β
Glass	<5
SiC, Al_2O_3, C	5–10
Steel	>100

$$\bar{\sigma} = \int_0^\infty \exp(-x) \frac{(x)^{\frac{\beta+1}{\beta}-1}}{(L\alpha)^{1/\beta}} dx \tag{10.6}$$

Now, letting $(\beta+1)/\beta=n$, we can convert Eq. (10.6) to the following simple form:

$$\bar{\sigma} = (L\alpha)^{-1/\beta} \int_0^\infty \exp(-x) x^{n-1} dx \tag{10.7}$$

The integral part of Eq. (10.7) is called the gamma function of order n and is denoted by $\Gamma(n)$, i.e.

$$\Gamma(n) = \int_0^\infty \exp(-x) x^{n-1} dx$$

Values of $\Gamma(n)$ can be found in mathematical handbook tables. Some important values of $\Gamma(n)$ are as follows:

$$\Gamma(0) = \infty$$
$$\Gamma(1/2) = \sqrt{\pi}$$
$$\Gamma(1) = 1$$
$$\Gamma(2) = 1$$
$$\Gamma(3) = 2!$$

Using the gamma function as defined above, we can write the mean or average strength of the fiber as

$$\bar{\sigma} = (L\alpha)^{-1/\beta} \Gamma(n) \tag{10.8}$$

Substituting the value of $n=(\beta+1)/\beta$, we have the mean strength given by

$$\bar{\sigma} = M_1 = (L\alpha)^{-1/\beta} \Gamma\left(1+\frac{1}{\beta}\right) \tag{10.9}$$

Now, the standard deviation s of a distribution is given by

$$s^2 = \mu_2 = M_2 - \bar{\sigma}^2$$

so that

$$s = \sqrt{\mu_2} = \sqrt{M_2 - M_1^2} \tag{10.10}$$

We can obtain an expression for s for the Weibull distribution under consideration by using the expressions for M_2 and M_1 involving $f(\sigma)$ as follows:

$$s=\left[\int_0^\infty \sigma^2 f(\sigma)\,\mathrm{d}(\sigma)-\left(\int_0^\infty \sigma f(\sigma)\,\mathrm{d}(\sigma)\right)^2\right]^{1/2} \qquad (10.11)$$

Let us evaluate the first integral in this expression. The second integral is given by Eq. (10.7). Substituting for $f(\sigma)$, we obtain for the first integral in Eq. (10.11)

$$\int_0^\infty L\alpha\beta\sigma^{(\beta-1)}\,\sigma^2\exp\left(-L\alpha\sigma^\beta\right)\mathrm{d}\sigma$$

$$=\int_0^\infty \exp\left(-L\alpha\sigma^\beta\right)\frac{(L\alpha\sigma^\beta)^{2/\beta}}{(L\alpha)^{2/\beta}}\,\mathrm{d}(L\alpha\sigma^\beta)$$

Denoting $L\alpha\sigma^\beta$ by x, we may rewrite the above integral as

$$\int_0^\infty \exp\left(-x\right)\frac{(x)^{\frac{\beta+2}{\beta}-1}}{(L\alpha)^{2/\beta}}\,\mathrm{d}x$$

$$=(\alpha L)^{-2/\beta}\,\Gamma\left(1+\frac{2}{\beta}\right)$$

Putting this gamma function form for the first integral in Eq. (10.11) and using the expression in Eq. (10.7) for the second integral, we obtain the following expression for the standard deviation, s:

$$s=\left[\int_0^\infty \sigma^2 L\alpha\beta\sigma^{\beta-1}\exp(-L\alpha\sigma^\beta)\,\mathrm{d}\sigma-\left((L\alpha)^{-1/\beta}\,\Gamma\left(1+\frac{1}{\beta}\right)\right)^2\right]^{1/2}$$

From the standard deviation and mean strength, we can arrive at the following expression for the coefficient of variation, μ, for this distribution

$$s=(\alpha L)^{-1/\beta}\left[\Gamma\left(1+\frac{2}{\beta}\right)-\Gamma^2\left(1+\frac{1}{\beta}\right)\right]^{1/2} \qquad (10.12)$$

$$\mu=\frac{s}{\bar\sigma}=\frac{\left[\Gamma\left(1+\frac{2}{\beta}\right)-\Gamma^2\left(1+\frac{1}{\beta}\right)\right]^{1/2}}{\Gamma\left(1+\frac{1}{\beta}\right)} \qquad (10.13)$$

We note from the above expression that μ is an inverse function of the Weibull modulus, β.

We can also find the statistical mode, σ^*, the most probable strength value for a given distribution. Consider the mean fiber strength (Eq. 10.8) again

$$\bar\sigma=(\alpha L)^{-1/\beta}\,\Gamma\left(1+\frac{1}{\beta}\right)$$

For a unit length of fiber, i.e. $L=1$, we can rewrite the above expression as

$$\bar{\sigma}=k\alpha^{-1/\beta}$$

where

$$k=\Gamma\left(1+\frac{1}{\beta}\right)$$

For $\beta>1$, we shall have $0.88 \leq k \leq 1.0$. This means that we can regard the quantity $\alpha^{-1/\beta}$ as the *reference level strength*. To find the statistical mode, σ^*, the most probable strength value, we proceed as follows. The Weibull distribution is

$$f(\sigma)=L\alpha\beta\sigma^{(\beta-1)}\exp(-L\alpha\sigma^{\beta})$$

At $\sigma=\sigma^*$ (mode value), the derivative of $f(\sigma)$ with respect to σ will be zero. We can write

$$\frac{\mathrm{d}f(\sigma)}{\mathrm{d}\sigma}=L\alpha\beta\,(\beta-1)\,\sigma^{(\beta-2)}\exp(-L\alpha\sigma^{\beta})$$

$$+L\alpha\beta\sigma^{(\beta-1)}\exp(-L\alpha\sigma^{\beta})\,[-L\alpha\beta\sigma^{(\beta-1)}]$$

$$=L\alpha\beta\sigma^{(\beta-1)}\left[\frac{(\beta-1)}{\sigma}\exp(-L\alpha\sigma^{\beta})-L\alpha\beta\sigma^{(\beta-1)}\exp(-L\alpha\sigma^{\beta})\right]$$

$$=L\alpha\beta\sigma^{(\beta-1)}\exp(-L\alpha\sigma^{\beta})\left[\frac{(\beta-1)}{\sigma}-L\alpha\beta\sigma^{(\beta-1)}\right]=0$$

so that

$$(\beta-1)=L\alpha\beta\sigma^{\beta}$$

Putting $\sigma=\sigma^*$, the mode value, we obtain

$$\sigma^*=(L\alpha)^{-1/\beta}\left(\frac{\beta-1}{\beta}\right)^{1/\beta} \tag{10.14}$$

For large β, we have

$$\frac{\beta-1}{\beta}\Rightarrow 1$$

and thus the mode strength can be written as

$$\sigma^*=(\alpha L)^{-1/\beta} \tag{10.15}$$

10.4   Experimental determination of Weibull parameters for a fiber

A statistical analysis of the fiber tensile strength values determined on a series of fiber samples can be easily made by using the two-parameter Weibull distribution described above. Using the form of the Weibull expression given in Eq. (10.2), we can write the probability of failure $F(\sigma)$ of the fiber at a stress σ, as

$$F(\sigma)=1-\exp\left(-\alpha\sigma^\beta\right) \qquad (10.16)$$

where β is the Weibull modulus and α is a scale parameter. As we said above, β is a measure of the scatter in the tensile data. Rearranging Eq. (10.16), we get

$$\ln\left[\ln\left(\frac{1}{1-F(\sigma)}\right)\right]=\beta\ln\sigma+\ln\alpha \qquad (10.17)$$

From Eq. (10.17), we can easily obtain α and β graphically. One generally arranges the tensile strength values of single filaments in ascending order and assigns a probability of failure using an estimator given by

$$F(\sigma_i)=\frac{i}{(1+N)} \qquad (10.18)$$

where $F(\sigma_i)$ is the probability of failure corresponding to the ith strength value and N is the total number of fibers tested. Substituting Eq. (10.18) in Eq. (10.17), we have

$$\ln\left[\ln\left(\frac{N+1}{N+1-i}\right)\right]=\beta\ln\sigma_i+\ln\alpha \qquad (10.19)$$

Equation (10.19) says then that a plot of $\ln\left[(N+1)/(N+1-i)\right]$ versus σ_i on a log–log graph would be a straight line if the tensile strength data followed a Weibull distribution. The intercept on the y-axis will then be α and the slope will be β. Figure 10.5a shows such a double log plot for the tensile strength of a mullite (Nextel 440) fiber, 12 mm gage length (Xu et al., 1993), while Fig. 10.5b shows a similar plot for carbon fiber (AS4), 10 mm gage length (Wu et al., 1991). Table 10.2 gives the Weibull modulus values for a series of fibers. These values should be taken as indicative values only. Similar use of Weibull two-parameter distribution has been made for other fibers, see for example Northolt and Sikkema (1991), Phani (1988) and Phoenix (1974a, 1974b). Although the two-parameter Weibull distribution is widely used, it does have some drawbacks. A significant discrepancy is observed between the values of Weibull modulus, also called the shape parameter, as obtained from a Weibull plot for a single gage length, L and from a $\log\sigma$ versus $\log L$ plot. In this regard, it is instructive to

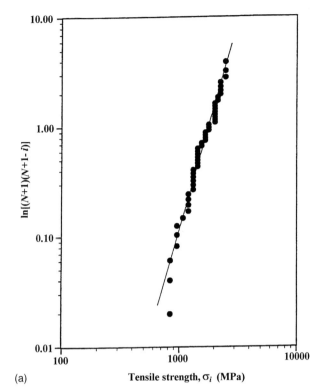

(a)

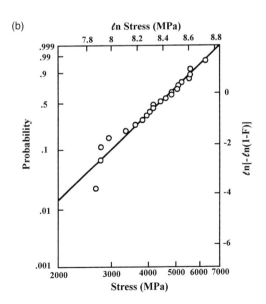

(b)

Figure 10.5 (a) Double log plot for the tensile strength of a mullite (Nextel 440) fiber, 12 mm gage length (after Xu *et al.*, 1995). (b) Double log plot for the tensile strength of a carbon fiber (AS4), 10 mm gage length (Wu *et al.*, 1991).

examine the extensive work of Wagner and colleagues (see Wagner *et al.*, 1989; Wagner, 1989) on the variability in strength of Kevlar aramid fiber. One of their conclusions was that Kevlar 49 filaments showed a significant variation in diameter within a yarn. This resulted in variabilty of fiber strength, i.e. failure load, but not in fiber tenacity or specific strength at failure. They observe an approximately linear correlation between the failure load and linear density, but no correlation was observed between linear density and specific strength or tenacity. This indicates the importance of spool to spool variability in a given fiber. Their plot of tenacity against gage length for Kevlar 49 showed a stronger fiber at smaller gage length, but the effect was less than that predicted by the Weibull model. The subject of variability in strength of a fiber really involves a complex interplay of statistical and material effects. For example, Göring *et al.* (1990) examined the effect of diameter variability in the case of carbon (T800) and Nicalon fibers. The statistical effect is that the strength decreases with increasing volume of fiber tested. The material effect is that fiber structure may vary with diameter. They eliminated the statistical effect by taking into account a unit volume of the fiber. Even after this correction, they observed a decrease in fiber strength with increasing fiber diameter, i.e. this effect was a true material effect.

The Weibull mean tensile strength, $\bar{\sigma}$, standard deviation, s, and coefficient of variation (CV) can be obtained

$$\bar{\sigma} = \alpha^{-1/\beta} \Gamma(1 + 1/\beta) \tag{10.20}$$

$$s = \alpha^{-1/\beta} [\Gamma(1 + 2/\beta) - \Gamma^2(1 + 1/\beta)]^{1/2} \tag{10.21}$$

where

$$\Gamma(n) = \int_0^\infty e^{-x} x^{n-1} dx$$

$$CV = 100 \left(\frac{s}{\bar{\sigma}} \right)$$

Table 10.2 *Typical Weibull modulus (β) for some fibers.*

Fiber	Glass	Carbon	Boron	Aramid	Ceramic[a]
Weibull modulus, β	10–12	5–6	3–6	10–12	3–6

Note:
[a] Fibers such as SiC, Al_2O_3, $(Al_2O_3) + SiO_2$).

Optical glass fibers

The telecommunication industry requires optical glass fibers with very high strength (>1 GPa) over very long lengths (>1 km). For example, the TrueWave single mode fiber of Lucent Technologies is routinely subjected to a proof stress of 0.7 GPa (see below) while the Weibull modulus is claimed to be 147 (Kummer, 1996). There is a variety of standards for measuring the tensile strength of optical fiber, for example, in the US the EIA/TIA-455-28B is the tensile test standard, due to the Electronic Industries Association and Telecommunication Industry Association, that is commonly used. It requires the data to be analyzed by Weibull statistics. As described in Chapter 7, the optical fiber is drawn in extremely clean environments, similar to the *clean rooms* associated with the semiconductor industry. Protective coating is applied to freshly drawn fiber which protects it from any environmental effects and also introduces compressive stresses on the surface. Pristine glass is a very high strength material, with a tensile strength greater than 3 GPa. It is the ever present moisture in the air and tiny impurities that lower the strength. Figure 7.13 in Chapter 7 shows the Weibull plots of optical fibers for two different gage lengths (Maurer, 1985). The Weibull modulus for 1 m gage length fiber is >50. The plot for the 10 m gage length sample, however, shows a bimodal character, indicating a multiple flaw population.

Here we should mention the proof testing technique, which is commonly used with optical glass fibers. Proof testing is a general technique in which a component is subjected for a short time to a stress, in excess of that expected in service. The idea is to weed out weak members from a set of components. The danger is that the proof testing may lead to some subcritical crack growth. In the case of an optical glass fiber, the fiber is taken off the spool and passed through a series of pulleys wherein it is first loaded in tension to a constant stress level much higher than that expected in service and then unloaded through another set of pulleys. All fibers containing flaws exceeding a certain size will fail in a proof test. Thus, this technique allows us to put an upper limit on the flaw size. This truncation of the flaw size distribution makes the surviving fibers stronger than the original set of fibers. The higher the proof strength, the more reliable will be the fiber but with the added penalty of a smaller fiber length. Recall that at a higher proof stress level, the fiber will break more often.

10.5 Fiber bundle strength

It is of interest to see how fibers behave when they are tested as a bundle in parallel. An untwisted yarn is a bundle of parallel fibers. As we shall presently see, it turns out that the strength of a bundle of fibers whose elements do not have a uniform strength is *less* than the average strength of the fibers. The fact that the mean strength of a fiber bundle is less than the mean strength of fibers tested individually can be visualized in a qualitative manner rather easily. When we load an untwisted yarn or bundle of fibers, breaks occur in fibers in a random manner. As a few fibers break, the load that was supported by them is thrown on to other unbroken fibers. Thus, the stress on the unbroken fibers becomes more than the average fiber stress. Finally, the bundle fails when the surviving fibers cannot support the applied load. This failure mechanism involving random fiber fracture leading to an overloading of the intact fibers is responsible for the average fiber bundle stress being less than the mean strength of fibers tested individually. Coleman (1958) treated this problem by making the following simplifying assumptions:

- All fibers are linear elastic;
- the fibers have an identical modulus and cross-sectional area;
- all individual filament strength is governed by Weibull distribution.

Following Daniels (1945) we use a normal distribution for the fiber bundle strength, σ_B. This is reasonable when the number of fibers, N, in a bundle is large. The mean value for a normal distribution is

$$\overline{\sigma}_B = \sigma_{fu}[1 - F(\sigma_{fu})] \tag{10.22}$$

where σ_{fu} is the maximum fiber strength corresponding to the condition of maximum load on the bundle. The maximum fiber strength, σ_{fu}, is obtained by equating the derivative of Eq. (10.22) to zero at $\sigma = \sigma_{fu}$. Thus,

$$\frac{d}{d\sigma}[\sigma(1 - F(\sigma))]_{\sigma = \sigma_{fu}} = 0$$

Normal distribution for the fiber bundle strength is

$$\omega(\sigma_B) = \frac{1}{\Psi_B \sqrt{2\pi}} \exp\left[-\frac{1}{2}\left(\frac{\sigma_B - \overline{\sigma}_B}{\Psi_B}\right)^2\right] \tag{10.23}$$

where Ψ_B, the standard deviation, is given by

$$\Psi_B = \sigma_{fu}[F(\sigma_{fu})(1 - F(\sigma_{fu}))]^{1/2} N^{-1/2} \tag{10.24}$$

and N is the number of fibers in the bundle. Note that as N increases, the standard deviation of the bundle strength, Ψ_B decreases.

Consider the Weibull distribution for the individual fiber Eq. (10.2). Then we can obtain σ_{fu} from

$$\frac{d}{d\sigma}[\sigma(1 - F(\sigma))]_{\sigma = \sigma_{fu}} = 0$$

where

$$F(\sigma) = \int_0^\sigma L\sigma B\sigma^{(\beta-1)} \exp(-L\alpha\sigma^\beta)\,d\sigma$$

Thus, we can write

$$\frac{d}{d\sigma}\left[\sigma\left(1 - \int_0^\sigma L\alpha\beta\sigma^{(\beta-1)}\exp(-L\alpha\sigma^\beta)\,d\sigma\right)\right]_{\sigma=\sigma_{fu}} = 0 \qquad (10.25)$$

Let us first evaluate the integral corresponding to $F(\sigma)$ in Eq. (10.25). Let $\sigma^\beta = x$, then $\beta\sigma^{(\beta-1)}d\sigma = dx$. The new limits of integration are: $\sigma = 0$, $x = 0$ and $\sigma = \sigma$, $x = x^{1/\beta}$.

$$F(\sigma) = \int_0^{x^{1/\beta}} L\alpha \exp(-L\alpha x)\,dx = L\alpha\left[\frac{\exp(-L\alpha x)}{-L\alpha}\right]_0^{x^{1/\beta}}$$

$$F(\sigma) = [-\exp(-L\alpha\sigma^\beta) + 1] \qquad (10.26)$$

Inserting Eq. (10.26) into Eq. (10.25), we get

$$\frac{d}{d\sigma}[\sigma\exp(-L\alpha\sigma^\beta)]_{\sigma=\sigma_{fu}} = 0$$

$$\exp(-L\alpha\sigma^\beta) + \sigma\exp(-L\alpha\sigma^\beta)(-L\alpha\beta\sigma^{\beta-1}) = 0$$

$$\exp(-L\alpha\sigma^\beta)[1 - (\sigma L\alpha\beta\sigma^{\beta-1})]_{\sigma=\sigma_{fu}} = 0$$

$$[1 - L\alpha\sigma^\beta]_{\sigma=\sigma_{fu}} = 0$$

$$\sigma_{fu} = \frac{1}{L\alpha\beta} \qquad (10.27)$$

$$\sigma_{fu} = [L\alpha\beta]^{-1/\beta} \qquad (10.28)$$

Therefore, from Eqs. (10.22) and (10.28), we obtain the mean fiber bundle strength as

$$\bar{\sigma}_B = (L\alpha\beta)^{-1/\beta}\exp(-L\alpha\sigma_{fu}^\beta) \qquad (10.8)$$

Now, from Eqs (10.27) and (10.29), we get

$$\bar{\sigma}_B = (L\alpha\beta)^{-1/\beta}\exp\left|-L\alpha\frac{1}{(L\alpha\beta)^{1/\beta}}\right|$$

$$\bar{\sigma}_B = (L\alpha\beta)^{-1/\beta}\exp\left(-\frac{1}{\beta}\right) \qquad (10.30)$$

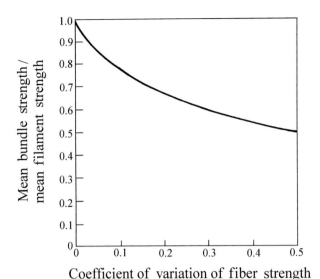

Figure 10.6   Fiber bundle strength/fiber strength as a function of coefficient of variation of fiber strength. The fiber bundle strength falls as the coefficient of variation of fiber strength increases.

Taking the ratio of the mean fiber bundle strength (Eq. (10.30)) and mean individual fiber strength (Eq. (10.9)), we can write

$$\frac{\sigma_B}{\overline{\sigma}_L} = \frac{(L\alpha\beta e)^{-1/\beta}}{(L\alpha)^{-1/\beta}\Gamma\left(1+\dfrac{1}{\beta}\right)}$$

$$= [\beta^{1/\beta}\exp(1/\beta)\Gamma(1+1/\beta]^{-1} \tag{10.31}$$

Here, we have implicitly assumed that the fiber bundle and the individual fibers have the same gage length. This ratio is a function of β, which, in turn, depends only on the coefficient of variation of the fiber. Figure 10.6 shows a plot of the ratio of mean fiber bundle strength to mean individual fiber strength as a function of coefficient of variation (μ) of the fiber. As expected, when the fibers show identical strength, coefficient of variation (μ) being zero, the ratio is unity. When the coefficient of variation of the fiber strength is 10%, the mean fiber bundle strength drops to 80% of the mean fiber strength. For a coefficient of variation of the fiber strength of 25%, the mean fiber bundle strength drops to 63% of the mean fiber strength.

References

Chapter 1

Chawla, K.K. (1987) *Composite Materials: Science & Engineering*, Springer-Verlag, New York.

Chawla, K.K. (1993) *Ceramic Matrix Composites*, Chapman & Hall, London.

Chou, T.-W. and F.K. Ko (eds) (1989) *Textile Structural Composites*, Elsevier, Amsterdam.

Clyne, T.W. and P.J. Withers (1993) *An Introduction to Metal Matrix Composites*, Cambridge University Press, Cambridge.

Hull, D. and T.W. Clyne (1996) *An Introduction to Composite Materials*, 2nd edn, Cambridge University Press, Cambridge.

Lyman, D.J. (1991) in *High-Tech Fibrous Materials*, ACS Symposium Series No. 457,

T.L. Vigo and A.F. Turbak (eds), American Chemical Society, Washington, DC.

New York Times (1993, July 13).

Piggott, M.R. (1980) *Load Bearing Fibre Reinforced Composites*, Pergamon Press, New York.

Sharp, A.G. (1995) *Invention and Technology*, Fall, 64.

Suresh, S., A. Needleman and A.M. Mortensen (eds) (1993) *Metal Matrix Composites*, Butterworth-Heinemann, Boston.

Taya, M. and R. J. Arsenault (1989) *Metal Matrix Composites*, Pergamon Press, New York.

Chapter 2

Anderson, K.J. (1993) *MRS Bull.*, August, 74.

Chawla, K.K. (1987) *Composite Materials: Science & Engineering*, Springer-Verlag, New York.

Chawla, K.K. (1993) *Ceramic Matrix Composites*, Chapman & Hall, London.

Chawla, K.K., Xu, Z.R., Hlinak, A. and Chung, Y.W. (1993) in *Advances in Ceramic Matrix Composites*, American Ceramic Society, Westerville, OH, p. 725.

Crook, L. (1993) in *Nonwovens: Theory, Process, Performance, and Testing*, A. Turbak (ed.), Tappi Press, Atlanta, p. 158.

Dresher, W.H. (1969) *J. Metals*, **21**, April, 17.

Ko, F.K. (1987) in *Engineered Materials Handbook*, Vol. 1, ASM International, Metals Park, OH, p. 519.

Lyman, D.J. (1991) in *High-Tech Fibrous Materials*, T.L. Vigo and A.F. Turbak (eds), American Chemical Society, Washington, DC, p. 116.

Mohamed, M.H. (1990) *American Scientist*, **78**, 530.

Perez, G. (1985) in *High Speed Fiber Spinning*, John Wiley, New York, p.333.

Vassilatos, G., Knox, B.H. and Frankfort, H.R.E. (1985) in *High Speed Fiber Spinning*, John Wiley, New York , p. 367.

Chapter 3

Barkakaty, B.C. *(1976) Journal of Applied Polymer Science*, **20**, 2921.

Batra, S.K. (1985) in *Fiber Chemistry, Handbook of Fiber Science and Technology: Volume IV*, M. Lewin and E.M. Pearce (eds), Marcel Dekker, New York, p. 727.

Berenbaum, M. (1995) *The Sciences*, **33**(5), 13.

Chand, N. and P.K. Rohatgi (1994) *Natural Fibers and Their Composites*, Periodical Experts Book Agency, Delhi, India.

Chawla, K.K. (1976) in *Proc. Int. Conf. Mech. Behavior of Materials II*, ASM, Metals Park, Ohio, p. 1920.

Chawla, K.K. and A.C. Bastos (1979) in *Proc. Int. Conf. Mech. Behavior of Materials III*, Pergamon Press, Oxford, p. 191.

Datye, K.R. and V.N. Gore (1994) *J. Geotextiles and Geomembranes*, **13**, 371.

Feltwell, J. (1990) *The Story of Silk*, St Martin's Press, New York.

Gosline, J.M., M.E. Demont and M.W. Denny (1986) *Endeavour*, New Series, **10**, January, 37.

Gosline, J.M., C. Nichols, P. Guerette, A. Cheng and S. Katz (1995) in *Biomimetics-Design and Processing of Materials*, M. Sarikaya and I.A. Aksay (eds), Amer. Inst. Physics Press, Woodbury, NY, p. 237.

Hyde, N. (1984) *National Geographic*, **165**, January, 2.

Jefferies, R., D.M Jones, J.K. Roberts, K. Selby, S.C. Simmons and J.O. Warwicker (1969) *Cell. Chem. Tech.*, **3**, 255.

Kaniraj, S.R. and G.V. Rao (1994) *J. Geotextiles and Geomembrane*, **13**, 389.

Kaplan, D., W.W. Adams, B. Farmer and C. Viney (eds) (1994) *Silk Polymers: Materials Science and Biotechnology*, American Chemical Society, Washington, DC.

Kerkam, K., C. Viney, D. Kaplan and S. Lombardi (1991) *Nature*, **349**, 596

Li, S.F.Y, A.J. McGhie and S.L. Tang (1994) *Biophysical Journal*, **66**, 1209.

Makinson, K.R. (1972) *Wool Sci. Rev.*, **43**, 2.

Meredith, R. (1970) *Contemporary Physics*, **11**, 43.

Robinson, R.M. (1985) in *Fiber Chemistry, Handbook of Fiber Science and Technology: Volume IV*, M. Lewin and E.M. Pearce (eds), Marcel Dekker, New York, p. 647.

Rohatgi, P.K., K.G. Satyanarayana and N. Chand (1992) in *International Encyclopedia of Composites*, S.M. Lee (ed), VCH, New York, **4**, p. 8.

Roe, P.J. and M.P. Ansell (1985) *J. Mater. Sci.*, **20**, 4015.

Segal, L. and P.H. Wakelyn (1985) in *Fiber Chemistry, Handbook of Fiber Science and Technology: Volume IV*, M. Lewin and E.M. Pearce (eds), Marcel Dekker, New York, p. 809.

Shear, W.A. (1994) *American Scientist*, **82**, 256.

Stout, H.P. (1985) in *Fiber Chemistry, Handbook of Fiber Science and Technology: Volume IV*, M. Lewin and E.M. Pearce (eds), Marcel Dekker, New York, p. 701.

Thiel, B.L. and C. Viney (1995) *MRS Bull.*, **20**, September, 54.

Thompson, J. (1994) *National Geographic*, June, 60.

Vollrath, F.(1992) *Scientific American*, **70**, March, 266.

Vollrath, F., W.J. Fairbrother, R.J.P. Williams, E.K. Tillinghast, D.T. Bernstein, K.S. Gallagher and M.A. Townley*, Nature*, **345**, 526.

Young, R.J. (1989) in *Handbook of Fiber Science and Technology*, Volume III, High Technology Fibers, M. Lewin and E.M. Pearce (eds), Marcel Dekker, New York.

Chapter 4

Allen, S.R. (1987) *J. Mater. Sci.*, **22**, 853.

Allen, S.R. (1988) *Polymer*, **29**, 1091.

Argon, A.S. (1972) in *Treatise on Materials Science and Technology*, Academic Press, New York, p. 79.

Bhattacharya, S. (1989) *Proc. Amer. Chem. Soc.*, Div. Poly. Mater. Sci. & Eng., **60**, 512.

Biro, D.A., G. Pleizier and Y. Deslandes (1992) *J. Mater. Sci. Lett.*, **11**, 698.

Brown, J.R., P.J.C. Chappell and Z. Mathys (1992) *J. Mater. Sci.*, **27**, 3167.

Bunn, C.W. (1939) *Trans. Faraday Soc.*, **35**, 482.

Capaccio, G., A.G. Gibson and I.M. Ward (1979) in *Ultra-High Modulus Polymers*, Elsevier, London, p. 1.

Couper, M. (1985) in *High Technology Fibers*, Part A, Part B, M. Lewin and J. Preston (eds), Marcel Dekker, New York, p. 113.

DeTeresa, S.J., S.R. Allen, R.J. Farris and R.S. Porter (1984) *J. Mater. Sci.*, **19**, 57.

DeTeresa, S.J., R.J. Farris and R.S. Porter (1982) *Polymer Composites*, **3**, 57.

DeTeresa, S.J., R.S. Porter and R.J. Farris (1985) *J. Mater. Sci.*, **20**, 1645.

DeTeresa, S.J., R.S. Porter and R.J. Farris (1988) *J. Mater. Sci.*, **23**, 1886.

Dobbs, M.G., D.J. Johnson and B.P. Saville (1980) *Phil. Trans. R. Soc. Lond.*, **A294**, 483.

Flory, P.J.(1956) *Proc. Roy. Soc. London*, **A234**, 73.

Greszczuk, L.B. *(1975) Amer. Inst. Aernautics and Astronautic Journal*, **3**, 1311.

Hadley, D.W., P.R. Pinnock and I.M. Ward (1969) *J. Mater. Sci.*, **4**, 152.

Hall, M.E. and A.E. Horrocks (1993) *Trends in Poymer Sci.*, **1**, 55.

Hild, D.N. and P. Schwartz (1992a) *J. Adhes. Sci. Technol.*, **6**, 879.

Hild, D.N. and P. Schwartz (1992b) *J. Adhes. Sci. Technol.*, **6**, 897.

Hodd, K.A. and D.C. Turley (1978) *Chemistry in Britain*, **14**, 545.

Hokudoh, T., K. Yabuki and Y. Nomura (1995) in *Proc. Second Int. Conf. on Composites Engineering*, New Orleans, LA, p. 333.

Holme, I. (1994) *J. Soc. Dyers. Color.*, **110**, 362.

Hounshell, D.A. and J.K. Smith (1988) *Science and Corporate Strategy*, Cambridge University Press, Cambridge.

Irwin, R.S. (1997) in *Applications of High Temperature Polymers*, R.R. Luise (ed), CRC Press, Boca Raton, FL, p. 149.

Jaffe, M. and R.S. Jones (1985) in *Handbook of Fiber Science & Technology*, Volume 3, High Technology Fibers, Part A, Marcel Dekker, New York, p. 349.

Jiang, H., Damodaran, S., Desai, P., Kumar, S. and Abhiraman, A.S. (1991) *Polym. Mater. Sci. Eng.*, **64**, 383.

Kalb, B. and A.J. Pennings (1980) *J. Mater. Sci.*, **15**, 2584.

Kaplan, S.L., P.W. Rose, H.X. Nguyen and H.W. Chang (1988) *SAMPE Quarterly*, **19**, 55.

Kikuchi, T (1982) *Surface*, **20**, 270.

Khosravi, N., S.B. Warner, N.S. Murthy and S. Kumar (1995) *J. Appl. Polym. Sci.*, **57**, 781.

Kozey, V.V. and S. Kumar (1994) *J. Mater. Res.*, **9**, 2717.

Kozey, V.V., H. Jiang, V.R. Mehta and S. Kumar (1994) *J. Mater. Sci.*, **10**, 1044.

Krassig, H.A. , J. Lenz and H.F. Mark (1984) *Fiber Technology : from Film to Fiber*, Marcel Dekker, New York.

Kumar, S. (1989), *SAMPE Quarterly*, **20**, 3.

Kumar, S., (1990a) in *International Encyclopedia of Composites 4*, VCH, Weinheim, p. 51.

Kumar, S. (1990b), *35th SAMPE Int. Symp.*, p. 2224.

Kumar, S. and W.W. Adams (1990), *Polymer*, **31**, 15.

Kumar, S., W.W. Adams and T.E. Heliminiack (1988) *J. Reinf. Plast. Compos.*, **7**, 108.

Kumar, S. and T.E. Heliminiack (1989a) in *The Materials Science and Engineering of Rigid-Rod Polymers*, MRS Symp., Vol. 134, Pittsburgh, p. 580.

Kumar, S. and T.E. Heliminiack (1989b) in *The Materials Science and Engineering of Rigid-Rod Polymers*, MRS Symp., Vol. 134, Pittsburgh, p. 363.

Li, Z.F., A.N. Netravali and W. Sachse (1992) *J. Mater. Sci.*, **27**, 4625.

Magat, E. E. (1980) *Phil. Trans. Roy. Soc. Lond.*, **A296**, 463.

Magat, E.E. and R.E. Morrison (1976) *Chemtech*, November, 702.

Mark, H.(1936) *Trans. Faraday Soc.*, **32**, 143.

Martin, D.C. and E.L. Thomas (1989) in *The Materials Science and Engineering of Rigid-Rod Polymers*, MRS Symp., Vol. 134, Mater. Res. Soc., Pittsburgh, p. 415.

McGarry, F.J. and J.E. Moalli (1991) *Polymer*, **32**, 35.

McGarry, F.J. and J.E. Moalli (1992) *SAMPE Quarterly*, July, 35.

Morgan, P. W. (1979) *Plastics and Rubber: Materials and Applications*, **4**, February, 1.

Nakajima, T. (ed.) (1994) *Advanced Fiber Spinning Technology*, Woodhead Pub. Co., Singapore.

Northolt, M.G. (1974) *Euro. Polymer Journal*, **10**, 799.

Northolt, M.G. (1981) *J. Materials Sci.*, **16**, 2025.

Ohta, T. (1983) *Polymer Eng. Sci.*, **23**, 697.

Ozawa, S., Y. Nakagawa, K. Matsuda, T. Nishihara and H. Yunoki (1978) US patent 4,075,172.

Panar, M., P. Avakian, R.C. Blume, K.H. Gardner, T.D. Gierke and H.H. Yang (1983) *Journal Polymer Sci., Polymer Phys.*, **21**, 1955.

Pennings, A.J., C.J.H. Schouten and A.M. Kiel (1972) *J. Polymer Sci.*, **C38**, 167 .

Pennings, A.J. (1976) *Colloid Polymer Sci.*, **253**, 452.

Polis, D.W., L.R. Dalton and D.J. Vachon (1989) *Materials Res. Soc. Symp. # 134*, Pittsburgh, p. 679.

Schuerch, H. (1966) *Amer. Inst. Aernautics and Astronautic Journal*, **4**, 102.

Smith P. and P.J. Lemstra (1976), *Colloid Polymer Sci.*, **15**, 258.

Smith P. and P.J. Lemstra (1980) *J. Mater. Sci.*, **15**, 505.

Spillman, G.E., L.J. Markoski, K.A. Walker, G.A Deeter, D.C. Martin and J.S. Moore (1993) *Polymeric Mater. Sci. and Eng.*, **69**, 139.

Smook, J. and A.J. Pennings (1984) *J. Mater. Sci.*, **19**, 31.

Swan, P.R. (1962) *J. Polymer Sci.*, **56**, 403.

van Stone, J.C. (1985) in *Handbook of Dialysis*, AXZO, Wuppertal, Germany, p. 21.

Wallenberger, F.T., N.E. Weston, K. Motzfeldt and D.G. Swartzfager (1992) *J. Am. Ceram. Soc.*, **75**, 629.

Wilson, N. (1967) *J. Text. Inst.*, **58**, 611.

Wilson, N. (1968) *J. Text. Inst.*, **59**, 296.

Wolfe, J.F., B.H. Loo and E.R. Seviller (1981a) *Polymer Preprint*, **22** (1), 60 .

Wolfe, J.F., B.H. Loo and F.E. Arnold (1981b) *Macromolecules*, **14**, 915 .

Wynne, K., A.E. Zachariades, T. Inabe and T.J. Marks (1985) *Polymer Commun.*, **26**, 162.

Young, R.J. (1989) in *Handbook of Fiber Science and Technology*, Volume 3, High Technology Fibers, Part A, Marcel Dekker, New York.

Chapter 5

Alber, N.E. and W.E. Smith (1965) US Patent 3 216 076.

Bewlay, B.P. (1991) in *Proc. 5th Int. Tungsten Symp.*, MPR Pub. Service, Shrewsbury, England, p. 227.

Bewlay, B.P., N. Lewis and K.A. Lou (1991) *Met. Trans.*, **22A**, 2153.

Briant, C.L. (1989) *Met. Trans.*, **20A**, 243.

Briant, C.L. and B.P. Bewlay (1995) *MRS Bulletin*, **20**, 67.

Donald, I.W. (1987) *J. Mater. Sci.*, **22**, 2661.

Embury, J.D. and R.M. Fisher (1966) *Acta Metall.*, **14**, Febuary, 147.

Engelke, J.L. (1967) US Patent 3 347 959.

Hall, E.O. (1951) *Proc. Phys. Soc. London*, **B64**, 747.

Hillman, H. (1981) in *Superconducting Materials Science*, S. Foner and B.B. Schwartz (eds), Plenum, New York, 275.

Horascek, O. (1989) *The Metallurgy of Doped/Non-Sag Tungsten*, Elsevier Applied Science, London, 251.

Hosford, W. and R.M. Caddell (1983) *Metal Forming*, Prentice-Hall, Englewood Cliffs, NJ.

Jaiswal, S. and I.D. McIvor (1989) *Ironmaking and Steelmaking*, **16**, 49.

Kikuchi, Y. and H. Shoji (1989) in *High Technology Fibers*, Part B, M. Lewin and J. Preston (eds), Marcel Dekker, New York, 175.

Meyers, M.A. and K.K Chawla (1984) *Mechanical Metallurgy*, Prentice-Hall, Englewood Cliffs, NJ, p. 353.

Oppelt, A. and T. Grandke (1993) *Supercond. Sci. Tech.*, **6**, 381.

Petch, N.J. (1953) *J. Iron Steel Inst.*, **174**, 25.

Pond, R.B. (1961) US Patent 2 976 590.

Taylor, G.F. (1924) *Phys. Rev.*, **23**, 655.

Takahashi, T., I. Ochiai and H. Satoh (1992) *Nippon Steel Tech. Report #53*, April, 102.

Vukcevich, N. (1991) in *Proc. 5th Int. Tungsten Symp.*, MPR Publishing Service Ltd., Shrewsbury, England, 157.

Wittenauer, J.P., T.G. Nieh, and J. Wadsworth (1992) *Adv. Mater. & Processes*, September, 28.

Chapter 6

Adler, R.P.I. and M.L. Hammond (1969) *Appl. Phys. Lett.*, **14**, 354.

Anon. (1993) *Bull. Am. Ceram. Soc.*, **72**, October 42.

Aveston, J. (1969) *J. Mater Sci.*, **4**, 625.

Birchall, J.D., J.A.A. Bradley and J. Dinwoodie (1985) in *Strong Fibres*, North-Holland, Amsterdam, 115.

Bogy, D.B. (1979) *Ann. Rev. of Fluid Mech.*, **11**, 207.

Brenner, S.S. (1958) in *Growth and Perfection of Crystals*, John Wiley, New York, p. 157.

Brenner, S.S. (1962) *J. Appl. Phys.*, **33**, 33.

Bunsell, A.R., G. Simon, Y. Abe and M. Akiyama (1988) in *Fibre Reinforcements for Composite Materials*, A.R. Bunsell (ed.), Elsevier, Amsterdam, p. 427 .

Chawla, K.K., (1993) *Ceramic Matrix Composites*, Chapman & Hall, London.

Chawla, N., J.W. Holmes and J.E. Mansfield (1995) *Materials Characterization*, **35**, 199.

Cooper, W.C. (1971) *Asbestos: The Need for and Feasibility of Air Pollution Controls*, Committee on Biological Effects of Atmosphere Pollutants, Div. of Medical Sciences/National Research Council, Washington, DC.

DeBolt, H.E. V.J. Krukonis and F.E. Wawner (1974) in *SIlicon Carbide 1973*, University of S. Carolina Press, Columbia, SC, p. 168.

Deren, G. (1995) *Bull. Amer. Ceram. Soc.*, **74**, 65.

Dhingra, A.K. (1980) *Philos. Trans. R. Soc. London*, **A294**, 411.

DiCarlo, J.A. (1985) *J. Metals*, **37**, June, 44.

DiCarlo, J.A. (1994) *Composites Science and Technology*, **51**, 213.

DiCarlo, J.A., H.M. Yun and J.C. Golsby (1995) in HITEMP Review, Vol. III, NASA Conf. Publication 10178, NASA Lewis Research Center, Cleveland, OH, p. 53.

Diefendorf, R.J. and L. Mazlout (1994) *Compos. Sci. and Technol.*, **51**, 181.

Economy, J. and R.V. Anderson (1967) *J. Polymer Sci.*, **C19**, 283.

Economy, J. and R. Lin (1977) in *Boron and Refractory Borides*, V.I. Matkovich (ed.), Springer-Verlag, Berlin, p. 552.

Fazen, P.J., J.S. Beck, A.T. Lynch, E.E. Remsen and L.G. Sneddon (1990) *Chem. Mater.*, **2**, 96.

Gasson, D.G. and B. Cockayne (1970) *J. Mater. Sci.*, **5**, 100.

Gooch, D.J. and G.W. Grover, (1973) *J. Mater. Sci.*, **8**, 1238.

Griffith, E.J. (1995) *Phosphate Fibers*, Plenum Press, New York,

Haggerty, J.S. (1972) *NASA-CR-120948*, NASA Lewis Res. Center, Cleveland, OH.

Hollar, Jr., W.E. and J.J. Kim (1991) *Ceram. Eng. Sci. Proc.*, **12**, 979.

Hurley, G.F. and J.T.A. Pollack (1972) *Met. Trans.*, **7**, 397.

Ishikawa, T. (1994) *Compos. Sci. Technol.*, **51**, 135.

Jakus, K. and V. Tulluri (1989) *Ceram. Eng. Sci. Proc.*, **10**, 1338.

Jang, T. and R.V. Subramanian (1993) *Scripta Met. Mater.*, **28**, 527.

Johnson, D.D., A.R. Holz and M.F. Grether (1987) *Ceram. Eng. Sci. Proc.*, **8**, 744.

Kimura Y., Y. Kubo and N. Hayashi (1994) *Compos. Sci. Technol.*, **51**, 173.

Krukonis, K. (1977) in *Boron and Refractory*

Borides, V.I. Matkovich (ed.), Springer-Verlag, Berlin, p. 517.

Kun, J., S. Tlali, H.E. Jackson, J.E. Webb and R.N. Singh (1996) *Appl. Phys. Lett.*, **68**, 2352.

LaBelle, H.E. and A.I. Mlavsky (1971) *Mater. Res. Bull.*, **6**, 571.

Laffon, C., A.M. Flank, P. Lagarde, M. Laridjani, R. Hagege, P. Olry, J. Cotteret, J. Dixmier, J.L. Niquel, H. Hommel and A.P. Legrand (1989) *J. Mater. Sci.*, **24**, 1503.

Laine, R.M. and F. Babonneau, (1993) *Chem. Mater.*, **5**, 260.

Laine, R.M., Z.-F. Zhang, K.W. Chew, M. Kannisto and C. Scotto (1995) in *Ceramic Processing Science and Technology*, Am. Ceram. Soc., Westerville, OH, p. 179.

Lara-Curzio, E. and S. Sternstein (1993) *Compos. Sci. Technol.*, **46**, 265.

Layden, J.K. (1973) *J. Mater. Sci.*, **8**, 1581.

Lee, J.-G. and I.B. Cutler (1975) *Am. Ceram. Soc. Bull.*, **54**, 195

Lesniewski C. , C. Aubin and A.R. Bunsell (1990) *Compos. Sci. & Technol.*, **37**, 63.

Lipowitz, J., T. Barnard, D. Bujalski, J. Rabe, G. Zank, A. Zangvil and Y. Xu (1994) *Compos. Sci. Technol.*, **51**, 167.

van Maaren, A.C., O. Schob and W. Westerveld (1975) *Phillips Tech. Rev.*, **35**, 125.

Mackenzie, K.J.D. and R.H. Meinhold (1994) *J. Mater. Sci.*, **29**, 2775.

Milewski, J.V., F.D. Gac, J.J. Petrovic and S.R. Skaggs (1985) *J. Mater. Sci.*, **20**, 1160.

Morscher, G.N. and H. Sayir (1995) *Mater. Sci. and Eng. A*, **190**, 267.

Narula, C.K., R. Schaffer, A.K. Datye, T.T. Borek, B.M. Rapko and R.T. Paine (1990) *Chem. Mater.*, **2**, 394.

Nourbakhsh, S., F.L. Liang and H. Margolin (1989) *J. Mater. Sci. Lett.*, **8**, 1252.

Okamura, K. and T. Seguchi (1992) *J. Inorganic and Organometallic Polymers*, **2**, 171.

Paciorek, K.J.L., Harris, D.H. and Kratzer, R.H. (1986) *J. Polym. Sci. Polym. Chem. Edn.*, **24**, 173.

Petrovic, J.J., J.V. Milewski, D.L. Rohr and F.D. Gac (1985) *J. Mater. Sci.*, **20**, 1167.

Pollack, J.T.A. (1972) *J. Mater. Sci.*, **7**, 787.

Pysher, D.J. and R.E. Tressler (1992) *J. Mater. Sci.*, **27**, 423.

Romine, J.C. (1987) *Ceram. Eng. Sci. Proc.*, **8**, 755.

Sacks, M.D., G.W. Scheiffele, M. Saleem, G.A. Staab, A.A. Morrone and T.J. Williams (1995) *Ceramic Matrix Composites: Advanced High-Temperature Structural Materials*, MRS, Pittsburgh, PA, p. 3.

Saitow, Y., K. Iwanaga and S. Itou (1992) *Proc. of the SAMPE Annual Meeting*, Vol. 37, Anaheim, CA.

Sayir, A. and A.C. Farmer (1995) *Ceramic Matrix Composites-Advanced High-Temperature Structural Materials*, MRS, Pittsburgh, PA, p. 11.

Sayir, A., A.C. Farmer, P.O. Dickerson and H.M. Yun (1995) *Ceramic Matrix Composites – Advanced High-Temperature Structural Materials*, MRS, Pittsburgh, PA, p. 21.

Schneider, H., K. Okada and J. Pask (1994) *Mullite and Mullite Ceramics*, John Wiley, Chichester.

Schneider, H., J. Göring, M. Schmücker and F. Flucht (1996) in *Ceramic Microstructures '96*, A.P. Tomsia and A. Glaeser (eds), Plenum Press, New York, in press.

Simon, G. and A.R. Bunsell (1984) *J. Mater. Sci.*, **19**, 3649.

Skinner, H.C.W., M. Ross, and C. Frondel (1988) *Asbestos and Other Fibrous Materials*, Oxford University Press, New York.

Smith, W.D. (1977) in *Boron and Refractory Borides*, V.I. Matkovich (ed.), Springer-Verlag, Berlin, p. 541.

Sowman, H.G. (1988) in *Sol-Gel Technology*, L.J. Klein (ed.), Noyes Pub., Park Ridge, NJ, p. 162.

Subramanian, R.V., H.F. Austin and T.J.Y. Wang (1977) *SAMPE Quarterly*, **8**, 1.

Talley, C.P. (1959) *J. Appl. Phys.*, **30**, 1114.

Toreki, W., C.D. Batich, M.D. Sacks, M. Saleem, G.J. Choi, and A.A. Morrone (1994) *Compos. Sci. Technol.*, **51**, 145.

Vega-Boggio, J. and O. Vingsbo (1978) in *1978 Int. Conf. Composite Materials*, ICCM/2, TMS-AIME, New York, p. 909.

Vega-Boggio, J. and O. Vingsbo (1976a) *J. Mater. Sci.*, **11**, 2242.

Vega-Boggio, J. and O. Vingsbo (1976b) *J. Mater. Sci.*, **11**, 2519.

Wallenberger, F.T., N.E. Weston, K. Motzfeldt and D.G. Swartzfager (1992) *J. Amer. Ceram. Soc.*, **75**, 629.

Wallenberger, F.T. and P.C. Nordine (1992) *Mater. Lett.*, **14**, 198.

Wawner, F.W. (1967) in *Modern Composite Materials*, L.J. Broutman and R.H. Krock (eds), Addison-Wesley, Reading MA, p. 244.

Wax, S.G. (1985) *Bull. Am. Ceram. Soc.*, **64**, 1096.

Weintraub, E. (1911) *J. Ind. Eng. Chem.*, **3**, 299.

Wills, R.R., R.A. Mankle and S.P. Mukherjee (1983) *Am. Ceram. Soc. Bull.*, **62**, 904.

Wilson, D.M. (1990) in *Proc. 14th Conf. On Metal Matrix, Carbon, and Ceramic Matrix Composites*, NASA Conf. Pub. 3097, Washington, DC, Part 1, p. 105.

Wilson, D.M., S.L. Lieder and D.C. Lueneburg (1995) *Ceram. Eng. Sci. Proc.*, **16**, 1005.

Xu, Z.R., K.K. Chawla and X. Li (1993) *Mater. Sci. and Eng.*, **A171**, 249.

Yajima, S., K. Okamura, J. Hayashi and M. Omori (1976) *J. Am. Ceram. Soc.*, **59**, 324.

Yajima, S. (1980) *Phil. Trans. R. Soc. London*, **A294**, 419.

Zarzycki, J. (1984) in *Glass: Science and Technology*, Vol. 2, D.R. Uhlmann and N. Kreidl (eds), Academic Press, New York, p. 209.

Zarzcyki, J., M. Prassas, and J. Phalippou (1982) *J. Mater. Sci.*, **17**, 3371.

Zhang, Z.-F., S. Scotto and R.M. Laine (1994) in *Ceram. Eng. Sci. Proc.*, **15**, 152.

Zhang, Z.-F., S. Scotto and R.M. Laine (1994) in *Covalent Ceramics II: Non-oxides*, MRS, Pittsburgh. PA, p. 207.

Chapter 7

Clare, A.G. (1995) in *Bioceramics: Materials and Applications*, Am. Ceram. Soc., Westerville, OH, p. 313.

Chandan, H.C., R.D. Parker and D. Kalish (1994) in *Fractography of Glass*, R.C. Bradt and R.E. Tressler (eds), Plenum Press, New York, p. 143.

Gossing, P. and G. Mahlke (1987) *Fiber Optic Cables*, John Wiley, New York.

Hannant, D.J. (1978) *Fibre Cements and Fibre Concretes*, John Wiley, New York.

Kenney, M.C., S.K. Barlow and S.L. Eikleberry (1996) *Nonwovens Conference*, TAPPI, Technology Park, Atlanta, GA, p. 149.

Majumdar, A.J. (1970) *Proc. Roy. Soc.*, **A319**, 69.

Maurer, R.D. (1985) in *Strength of Inorganic Glass*, C.R. Kurkjian (ed), Plenum, New York, p. 291.

Mecholsky, J.J., S.W. Freiman and S.N. Morey (1977) *Bull. Amer. Ceramic Soc.*, **56**, 1016.

Mecholsky, J.J., S.W. Freiman and S.N. Morey

(1979) in *Fiber Optics: Advances in Research and Development*, B. Bendow and S.S. Mitra (eds), Plenum, New York, p. 187.

Midwinter, J.E. (1979) *Optical Fibers for Transmission*, John Wiley, New York.

Nagel, S.R., J.B. MacChesney and K.L. Walker (1982) *IEEE J. Quant. Electron.*, **QE-18**, 459.

Proctor, B.A. (1971) *Composites*, **2**, 85.

Sakka, S. (1982) in *Treatise on Materials Science and Technology*, **22**, M. Tomozawa and R.H. Doremus (eds), Academic Press, New York, p. 129.

Scherer, G.W. (1980) *Applied Optics*, **19**, 2000.

Zarzycki, J. (1984) in *Glass: Science and Technology*, Vol. 2, D.R. Uhlmann and N. Kreidl (eds), Academic Press, New York, p. 209.

Zarzycki, J., M. Prassas and J. Phalippou (1982) *J. Mater. Sci.*, **17**, 3371.

Chapter 8

Baker, R.T.K. and P.S. Harris (1978) in *Chemistry and Physics of Carbon*, P.L. Walker and P.A. Thrower (eds), Vol. **14**, Marcel Dekker, New York, p. 83.

Bennett, S.C., D.J. Johnson and W. Johnson (1983) *J. Mater. Sci.*, **18**, 3337.

Deurbergue, A. and A. Oberlin (1991) *Carbon*, **29**, 621.

Edie, D.D. (1990) in *Carbon Fibers and Filaments*, J.L. Figueiredo, C.A. Bernardo, R.T.K. Baker and K.J. Hüttinger (eds), Kluwer Academic Publishers, Boston, p. 43.

Edie, D.D. and M.G. Dunham (1989) *Carbon*, **27**, 477.

Herakovich, C.T. (1989) *Carbon*, **27**, 663.

Hüttinger, K.J. (1990) *Adv. Mater.*, **2**, 349

Inal, O.T., N. Leca and L. Keller (1980) *Phys. Status Solidi*, **62**, 681.

Jain, M.K. and A.S. Abhiraman (1987), *J. Mater. Sci.*, **22**, 278.

Johnson, D.J. (1987) in *Chemistry and Physics of Carbon*, Vol. 20, P.A. Thrower (ed.), Marcel Dekker, New York, p. 1.

Johnson, D.J.(1990) in *Carbon Fibers and Filaments*, J.L. Figueiredo, C.A. Bernardo, R.T.K. Baker and K.J. Hüttinger (eds), Kluwer Academic Publishers, Boston, p. 119.

Kelly, B.T. (1981) *Physics of Graphite*, Applied Science Publishers, London.

Kowalska, M. (1982) *SAMPE Quarterly*, **13**, April, 2.

Kumar, S. (1989) *SAMPE Quarterly*, **20**, January, 3.

Meyers, M.A. and K.K Chawla (1984) *Mechanical Metallurgy*, Prentice-Hall, Englewood Cliffs, NJ, p. 188.

National Aeronautics and Space Administration

NASA (1978, 1979, 1980) Annual Reports, Washington, DC.

Oberlin, A., M. Endo and T. Koyama (1976) *J. Crystal Growth*, **32**, 335.

Piraux, L., B. Nysten, A. Haquenne, J.P. Issi, M.S. Dresselhaus and M. Endo (1984) *Solid State Commun.*, **50**, 697.

Robinson, K.E. and D.D. Edie (1996) *Carbon*, **34**, 13.

Reynolds, W.N. (1981) *Phil. Trans. Roy. Soc.*, London, **A29**, 451.

Rhee, B.S. (1990) *High Temperatures-High Pressures*, **22**, 267.

Riggs, J.P. (1985) in *Encyclopedia of Polymer Science and Engineering*, 2nd edn, Vol. **2**, John Wiley, New York, p. 640.

Singer, L.S. (1979) in *Ultra-High Modulus Polymers*, Applied Science Publishers, Essex, UK, p. 251.

Tibbetts, G.G. (1989) *Carbon*, **27**, 745.

Tibbetts, G.G., M. Endo and C. Beetz, Jr. (1986) *SAMPE J.*, **30**, 22.

Wagner, F.N. (1996) personal communication.

Watt, W. (1970) *Proc. Roy. Soc.*, **A319**, 5.

Chapter 9

Bascom, W.D. (1992) in *Modern Approaches to Wettability: Theory and Applications*, M.E. Schrader and G. Loeb (eds) Plenum Press, New York, p. 73.

Batchelder, D.N, C. Cheng and G.D. Pitt (1991) *Adv. Materials*, **3**, 566.

Berger, L.P., C.G. Plummer and H.H. Kausch, to be published.

Boogh, L.C.N., R.J. Meier, H.H. Kausch and B.J. Kip (1992) *J. Polymer Science, Physics*, **30**, 325.

Cannon, WR and E. Mendoza (1991) in *Proceedings. of the 4th International. Symposium. on Ceramic Mater. and Components for Engineers*, Gotheberg, Sweden.

Day, R.J. and R.J. Young (1993) *J. Microscopy*, **169**, 155.

Galiotis, C. (1991) *Compos. Sci. Technol.*, **42**, 125.

Greenwood, J.H. and P.G. Rose (1974) *J. Mater. Sci.*, **9**, 1809.

Hadley, D.W., P.R. Pinnock and I.M. Ward (1969) *J. Mater. Sci.*, **4**, 152.

Hagege, R. and A.R. Bunsell (1988) in *Fibre Reinforcement for Composite Materials*, A.R. Bunsell (ed.), Elsevier, London, p. 479.

Hillig, W.B. (1981) *Bull. Amer. Ceram. Soc.*, **66**, 373.

Hüttinger, K.J. (1990) *Adv. Mater.*, **2**, 349.

Jain, M.K. and A.S. Abhiraman (1987) *J. Mater. Sci.*, **22**, 278.

Jain, M.K., M. Balasubramanian, P. Desai and A.S. Abhiraman (1987) *J. Mater. Sci.*, **22**, 301.

Jawad, S.A. and I.M. Ward (1978) *J. Mater. Sci.*, **13**, 1381.

Jones, M.-C.G., E. Lara-Curzio, A. Kopper and D.C. Martin (1997) *J. Mater. Sci.*, **32**, 2855.

Kawabata, S. (1990) *J. Text. Inst.*, **81**, 432.

Kelsey, R.H. (1970) in *Whisker Technology*, Wiley-Interscience, p. 140.

Kip, B.J., M.C.P. van Eijk and R.J. Meier (1991) *J. Polymer Sci. Phy. Edn.*, **29**, 99.

Marsh, D.M. (1961) *J. Sci. Inst.*, **38**, 229.

McMahon, P.E. (1973) ASTM Special Technical Publicatioan 521, ASTM, Philadelphia, p. 357.

Mehta, V.R. (1996) PhD Thesis, Georgia Tech, Atlanta, GA.

Neumann A.W. and Good, R.J. (1979) *Surface and Colloid Science*, **11**, 31.

Nunes, J. and W. Klein (1967) *Trans. ASM*, **60**, 726.

Petrovic, J.J., J.V. Milewski, D.L. Rohr and F.D. Gac (1985) *J. Mater. Sci.*, **20**, 1167.

Phoenix, S.L. and J. Skelton (1974) **32**, *Res. Journal*, 934

Schultz, J. and M. Nardin (1992) in *Modern Approaches to Wettability: Theory and Applications*, M.E. Schrader and G. Loeb (eds), Plenum Press, New York, 359.

Sheaffer, P.M. (1987) in *Extended Abstracts of the 18th Biennial Conference on Carbon*, Worcester, MA, Amer. Carbon Society, Unviersity Park, PA, p. 20.

Siemers, P.A., R.L. Mehan and H. Moran (1988) *J. Mater. Sci.*, **23**, 329.

Sinclair, D. (1950) *J. Appl. Phys.*, **21**, 380.

Smith, R.E. (1972) *J. Appl. Phys.*, **43**, 255.

Sudiwala, R.V., C. Cheng, E.G. Wilson and D.N. Batchelder (1992) *Thin Solid Films*, **210**, 452.

Tsai, C.-L. and I.M. Daniel (1994) *Compos. Sci. Technol.*, **50**, 7.

Unal, D. and K.P. Lagerlof (1993) *J. Amer. Ceram. Soc.*, **76**, 3167.

Williams K.P.J., G.D. Pitt, D.N. Batchelder and B.J. Kip (1994) *Applied Spectroscopy*, **48**, 232.

Whittaker, A.J., M.L. Allitt, D.G. Onn and J.D. Bolt (1990) in *Thermal Conductivity 21*, C.J. Cremers and H.A. Fine (eds), Plenum Press, New York, p. 187.

Wong, W.F. and R.J. Young (1994) *J. Mater. Sci.*, **29**, 510.

Xu, Z.R., K.K. Chawla and X. Li (1993), *Mater. Sci. Eng.*, **A171**, 249.

Young, R.J., R.J. Day and M. Zakikhani (1990) *J. Mater. Sci.*, **25**, 127.

Young, R.J. (1996) in *Polymer Spectroscpy*, A.H. Fawcett (ed.), John Wiley, New York, p. 203.

Chapter 10

Coleman, B.D. (1958) *J. of the Mechanics and Physics of Solids*, **7**, 60.

Daniels, H.E. (1945) *Proc. Roy. Soc.*, **138A**, 405.

Diefendorf, R.J. and E. Tokarsky (1975) *Polymer Eng. Sci.*, **15**, 150.

Eldridge, J.I., J.P. Wiening, T.S. Davison and M.-J. Pindera (1993) *J. Am. Ceram. Soc.*, **7**, 3151.

Göring, J., F. Flucht and G. Ziegler (1990) in *4th European Conference on Composite Materials*, Elsevier, London, p. 543.

Herring, H.W. (1965) NASA TN D-3202, National Aeronautics and Space Administration, Langley, VA.

Kummer, R.B. (1996) personal communication.

Maurer, R.D (1985) in *Strength of Inorganic Glass*, C.R. Kurkjian (ed.), Plenum Press, New York, p. 291.

Northolt, M.G. and D.J. Sikkema (1991) in *Advances in Polymer Science 98*, Springer-Verlag, Berlin, p. 115.

Phani, K.K. (1988) *J. Mater. Sci.*, **23**, 1189.

Phoenix, S.L. (1974a) *Fibre Science and Technology*, **7**, 15.

Phoenix, S.L. (1974b) in *Composite Materials: Testing and Design*, ASTM, Philadelphia, p. 130.

Rosen, B.W. (1965) in *Fiber Composite Materials*, ASM, Metals Park, OH, p. 58.

Wagner, H.D. (1989) in *Application of Fracture Mechanics to Composite Materials*, Elsevier, Amsterdam, p. 39.

Wagner, H.D., S.L. Phoenix and P. Schwartz. (1989) *J. Compos. Mater.*, **18**, 312.

Weibull, W. (1951) *J. Appl. Mech.*, **18**, 293.

Wu, H.F., G. Biresaw and J.T. Laemmle (1991) *Polymer Compos.*, **12**, 281.

Xu, Z.R., K.K. Chawla and X. Li (1993) *Mater. Sci. Eng.*, **A171**, 249.

Zweben, C. (1977) *J. Mater. Sci.*, **7**, 1325.

Suggested Further Reading

Chapter 1

Baer, E., L.J. Gathercole and A. Keller (1975) in *Structures of Fibrous Proteins*, E.D.T. Atkins and A. Keller (eds), Butterworth, London, p. 189.

Bunsell, A.R. (ed.) (1988) *Fibre Reinforcements for Composites*, Elsevier, Amsterdam.

Hongu, T. and G.O. Phillips (1990) *New Fibers*, Ellis Horwood, New York.

Kostikov, V.I. (ed.) (1995) *Fibre Science Technology*, Chapman & Hall, London.

Morton, W.E. and J.W.S. Hearle (1993) *Physical Properties of Textile Fibers*, 3rd edn, The Textile Institute, London.

Warner, S.B. (1995) *Fiber Science*, Prentice Hall, Englewood Cliffs, NJ.

Watt, W. and B.V. Perov (eds) (1985) *Strong Fibres*, North-Holland, Amsterdam.

Chapter 2

Hongu, T. and G.O. Phillips (1990) *New Fibers*, Ellis Horwood, New York.

Moncrieff, R.W. (1975) *Man-Made Fibres*, Newnes Butterworths, London.

Morton, W.E. and J.W.S. Hearle (1993) *Physical Properties of Textile Fibers*, 3rd edn, The Textile Institute, London.

Warner, S.B. (1995) *Fiber Science*, Prentice Hall, Englewood Cliffs, NJ.

Chapter 3

Carter, M.E. (1971) *Essential Fiber Chemistry*, Marcel Dekker, New York.

Kaplan, D., W.W. Adams, B. Farmer and C. Viney (eds) (1994) *Silk Polymers: Materials Science and Biotechnology*, Amer. Chem. Soc., Washington, DC.

Kirby, R.H. (1963) *Vegetable Fibres*, Interscience, NY.

Chapter 4

Adams, W.W., R.K. Eby and D.E. McLemore (eds) (1989) *Materials Research Society Symp. 134*, MRS, Pittsburgh, PA.

Baer, E. and A.Moet (eds) (1991) *High Performance Polymers*, Hanser, Munich.

Carter, M. E. (1971) *Essential Fiber Chemistry*, Marcel Dekker, New York.

Crist, B. (1995) *Annual Rev. Mater. Sci.*, **25**, 295.

Donald, A.M. and A.H. Windle (1992) *Liquid Crystalline Polymers*, Cambridge University Press, Cambridge.

Jiang H., W.W. Adams and R.K. Eby (1993) in

Structure and Properties of Polymers, E.L. Thomas (ed.) VCH, Weinheim, Germany.

Reimschuessel, H. (1985) in *Fiber Chemistry, Handbook of Fiber Science and Technology: Volume IV*, M. Lewin and E.M. Pearce (eds), Marcel Dekker, New York, p. 73.

Warner, S.B. (1995) *Fiber Science*, Prentice Hall, Englewood Cliffs, NJ.

Watt, W. and B.V. Perov (eds) (1985) *Strong Fibres*, North-Holland, Amsterdam.

Yang, H.H. (1993) *Kevlar Aramid Fiber*, John Wiley, Chichester, UK.

Chapter 5

Donald, I.W. (1987) Production, properties and applications of microwire and related products, *J. Mater. Sci.*, **22**, 2661.

Pink, E. and L. Bartha (eds) (1989) *The Metallurgy of Doped/Non-Sag Tungsten*, Elsevier Applied Science, London.

Chapter 6

Brinker, C.J. and G. Scherer (1990) *The Sol–Gel Science*, Academic Press, New York.

Jones, R.W. (1989) *Fundamental Principles of Sol–Gel Technology*, The Institute of Metals, London.

Kingery, W.D., H.K. Bowen and D.R. Uhlmann (1976) *Introduction to Ceramics*, 2nd edn, John Wiley, New York.

Watt, W. and B.V. Perov (eds) (1985) *Strong Fibers*, North-Holland, Amsterdam.

Chapter 7

Kao, C.K. (1988) *Optical Fibre*, Peter Peregrinus, London.

Kao, C.K. (1982) *Optical Fiber Systems, Technology, Design and Applications*, McGraw-Hill, New York.

Li, T. (ed.) (1985) *Fiber Fabrication, Vol. 1 in the series: Optical Fiber Communications*,

Academic Press, Orlando, FL.

Murata, H. (1988) *Handbook of Optical Fibers and Cables*, Marcel Dekker, New York.

Rawson, H. (1990) *Glasses and Their Applications*, The Institute of Metals, London.

Scholze, H. (1991) *Glass*, Springer-Verlag, New York.

Chapter 8

Bunsell, A.R. (ed.) (1988) *Fibre Reinforcements for Composite Materials*, Elsevier, Amsterdam.

Chung, D.D.L.(1994) *Carbon Fiber Composites*, Butterworth-Heinemann, Boston.

Donnet, J.-B. and R.C. Bansal (1984) *Carbon Fibers*, 2nd edn, Marcel Dekker, New York.

Dresselhaus, M.S., G. Dresselhaus, K. Sugihara, I.L. Spain and H.A. Goldberg (1988) *Graphite*

Fibers and Filaments, Springer Series in Materials Science, Vol. 5, Springer-Verlag, New York.

Figueirdo, J.L., C.A. Bernardo, R.T.K. Baker and K.J. Hüttinger (eds) (1990) *Carbon Fibers and*

Filaments, Kluwer Academic, Boston.

Sittig, M. (1980) *Carbon & Graphite Fibers*, Noyes Data Corp., Park Ridge, NJ.

Peebles, L. H. (1995) *Carbon Fibers*, CRC Press, Boca Raton, FL.

Chapter 10

Kittl P. and G. Diaz (1988) *Res. Mechanica*, **24**, 99.

Zweben, C. (1977) *J. Mater. Sci.*, **7**, 1325

Appendix A

Some important units and conversion factors

A.1   Unit abbreviations

A=ampere
Å=angstrom
°C=degrees Celsius
cal=calorie
cm=centimeter
eV=electron volt
°F=degrees Fahrenheit
ft=foot
g=gram
GPa=gigapascal
in=inch
J=joule
K=kelvin
kg=kilogram
lb=pound force
lb=pound mass
m=meter
mm=millimeter
mol=mole
MPa=megapascal
N=newton
nm=nanometer

P=poise
Pa=pascal
s=second
μm=micrometer
W=watt
psi=pounds per square inch

A.2   SI multiple and submultiple prefixes

Factor by which multiplied	Prefix	Symbol
10^9	giga	G
10^6	mega	M
10^3	kilo	k
10^{-2}	centi	c
10^{-3}	milli	m
10^{-6}	micro	μ
10^{-9}	nano	n
10^{-12}	pico	p

A.3   Values of selected physical constants

Constant	Symbol	Value
Boltzmann's constant	k	1.38×10^{-23} J/mol K
Avogadro's number	N_A	6.023×10^{23} atoms or molecules/mol
Gas constant	R	8.31 J/mol K
Planck's constant	h	6.626×10^{-34} J s

A.4   Stress (or pressure)

1 dyne=10^5 newton (N)
$1\,N\,m^{-2}=10\,dyn\,cm^{-2}=1$ pascal (Pa)

$1\,\mathrm{MPA}=10^6\,\mathrm{N\,m^{-2}}=1\,\mathrm{N\,mm^{-2}}$
$1\,\mathrm{bar}=10^5\,\mathrm{N\,m^{-2}}=10^5\,\mathrm{Pa}$
$1\,\mathrm{kilobar}=10^8\,\mathrm{N\,m^{-2}}=10^9\,\mathrm{dyn\,cm^{-2}}$
$1\,\mathrm{mm\,Hg}=1\,\mathrm{torr}=133.322\,\mathrm{Pa}=133.32\,\mathrm{N\,m^{-2}}$
$1\,\mathrm{kgf\,mm^{-2}}=9806.65\,\mathrm{kN\,m^{-2}}=9806.65\,\mathrm{kPa}=100\ \mathrm{atmospheres}$
$1\,\mathrm{kgf\,cm^{-2}}=98.0665\,\mathrm{kPa}=1\ \mathrm{atmosphere}$
$1\,\mathrm{lb\,in^{-2}}=6.89476\,\mathrm{kN\,m^{-2}}=6.89476\,\mathrm{kPa}$
$1\,\mathrm{kgf\,cm^{-2}}=14.2233\,\mathrm{lb\,in^{-2}}$
$10^6\,\mathrm{psi}=10^6\,\mathrm{lb\,in^{-2}}=6.89476\,\mathrm{Gn\,m^{-2}}=6.89476\,\mathrm{GPa}$
$1\,\mathrm{GPa}=145\,000\,\mathrm{psi}$

A.5   Viscosity

$1\,\mathrm{poise}=0.1\,\mathrm{Pa\,s}=0.1\,\mathrm{Nm^{-2}s}$
$1\,\mathrm{GN\,m^{-2}s}=10^{10}\,\mathrm{poise}$

A.6   Energy per unit area

$1\,\mathrm{erg\,cm^{-2}}=1\,\mathrm{mJ\,m^{-2}}=10^{-3}\,\mathrm{Jm^{-2}}=1\,\mathrm{dyn\,cm^{-1}}=10^{-3}\,\mathrm{N\,m^{-1}}$
$10^8\,\mathrm{erg\,cm^{-2}}=47.68\,\mathrm{ft\,lb\,in^{-2}}=572.16\,\mathrm{psi\,in}$

A.7   Fracture toughness

$1\,\mathrm{psi\,in^{1/2}}=1\,\mathrm{lbf\,in^{-3/2}}=1.11\,\mathrm{kN\,m^{-3/2}}=1.11\,\mathrm{kPa\,m^{-1/2}}$
$1\,\mathrm{ksi\,in^{1/2}}=1.11\,\mathrm{MPa\,m^{-1/2}}$
$1\,\mathrm{MPa\,m^{1/2}}=0.90\,\mathrm{ksi\,in^{1/2}}$
$1\,\mathrm{kgf\,mm^{-2}mm^{1/2}}=3.16\times10^4\,\mathrm{N\,m^{-2}m^{1/2}}$

A.8   Density

$1\,\mathrm{g\,cm^{-3}}=62.4280\,\mathrm{lb\,ft^{-3}}=0.0361\,\mathrm{lb\,in^{-3}}$
$1\,\mathrm{lb\,in^{-3}}=27.68\,\mathrm{g\,cm^{-3}}$
$1\,\mathrm{g\,cm^{-3}}=10^3\,\mathrm{kg\,m^{-3}}$

A.9   Textile units

Linear density

$\mathrm{tex}=\mathrm{g/km}$
$\mathrm{denier}=\mathrm{g/9\,km}$

Specific stress

$1\,\mathrm{N/Tex} = 10^{-6}\,\mathrm{Pa}/\rho$

where ρ is density in $\mathrm{kg\,m^{-3}}$

$1\,\mathrm{N\,tex^{-1}} = 1\,\mathrm{GPa}/\rho$

where ρ is density in $(\mathrm{g\,cm^{-3}})$

A.10 Unit abbreviations

	g/denier	g/tex	kg/mm²	Pa	lb/in²
g/denier	1	0.111	$\rho\,8.995$	$\rho\,8.818\times10^7$	$\rho\,1.281\times10^4$
g/tex	9.000	1	$\rho\,80.96$	$\rho\,7.936\times10^8$	$\rho\,1.152\times10^5$
kg/mm²	$0.111/1\rho$	$0.0124/\rho$	1	9.803×10^7	1.281×10^4
Pa	$(1.138\times10^{-8})/\rho$	$(1.260\times10^{-9})/\rho$	1.019×10^{-8}	1	1.451×10^{-4}
lb/in²	$(7.013\times10^{-5})/\rho$	$(8.681\times10^{-6})/\rho$	7.013×10^{-4}	6.889×10^3	1

Author index

Subject index